A McColville

FYNBOS
PLANTS

A PRACTICAL GUIDE TO THE
PROPAGATION AND CULTIVATION OF PLANTS
FROM SOME OF THE MAJOR FAMILIES OF
THE CAPE FLORISTIC REGION OF SOUTH AFRICA

By Neville Brown and Graham Duncan
Photographs by Graham Duncan, Neville Brown and from the
Kirstenbosch Collection

Leucospermum cordifolium in the Protea Garden at Kirstenbosch

FOREWORD

It is now over ninety-three years since the founding of the National Botanical Gardens of South Africa at Kirstenbosch. This precursor to the organisation presently known as the S. A. National Biodiversity Institute had among its principal objectives, programmes to investigate the horticultural requirements of the indigenous flora of South Africa. The very notion of growing native flora scarcely existed among the gardening public in 1913. Moreover, for those then far-sighted enough to want to grow indigenous plants in their gardens, there was simply no published information at all. Indeed, with a flora of some 22 000 species to choose from of which at least 9000 are regarded as elements of fynbos it's not surprising that even the most dedicated gardener tended to feel overwhelmed for choice. Much has been learnt however, since those early days and in this excellent concise publication, Neville Brown and Graham Duncan have distilled the cumulative wisdom of decades of horticultural experience in growing fynbos.

Not all fynbos plants are easy to cultivate. Some of the most decorative and ornamental, especially those from high altitude montane habitats are almost impossible to grow in the average home garden, inevitably giving rise to disappointment. Here then is a book to help the beginner and even the more knowledgeable enthusiast, to navigate their way through the pitfalls of species choice. Based on nearly a century of trial and error as well as much recent smoke seed germination research at Kirstenbosch, it enumerates those fynbos species with a proven record as successful garden subjects, with specific directions on how to propagate and cultivate them.

With waterwise gardening uppermost in the minds of responsible gardeners this essentially practical guide should go a long way towards promoting the greening of urban environments while reducing domestic water consumption, especially in the winter rainfall area. The Botanical Society of South Africa has since its inception made tremendous efforts to encourage gardening with native, waterwise plants. It is our sincere hope that such an environmentally sound approach will gather more converts. May this informative and authoritative book be the inspiration towards achieving that goal.

DR JOHN ROURKE

President, Botanical Society of South Africa, Former Head of Compton Herbarium, SANBI

Oxalis obtusa flowering *en masse* near Villiersdorp

Dedicated to the memory of

Katherine Jeffcott Moody Bergh

1928 - 2005

In recognition of her tremendous contribution to plant conservation, which she made with great style and charm.

On arriving in South Africa from her hometown of Northfield, Massachusetts, USA, after her marriage to Cecil Bergh, Kay rose to the challenge of being a farmer's wife in the Cederberg Mountains and fell in love with the beauty of this country, its people and its wild flowers. Her role in plant conservation covered the full spectrum of flower shows (including Flora '83, '88 and '93), establishing the Ramskop Wildflower Garden in Clanwilliam, chairing the Botanical Society Council (the first lady to do so in the Society's history), serving on the Board of Trustees of the NBG (later NBI, now SANBI), establishing and driving the Botanical Society's wildflower guide publications committee, chairing the Kirstenbosch Development Campaign and making regular personal donations to Kirstenbosch and other conservation projects.

Prof. Brian Huntley, Chief Executive Officer, SANBI. In Memoriam, Veld & Flora, *March 2005*

Kay gave much to the Botanical Society but most of all she gave that most precious commodity of all - her time. She was a wonderful source of encouragement and inspiration. She had a great ability to motivate those around her. She displayed an exceptional talent for gathering around her a group of people or organizations that would ensure the success of a cause. She was unsurpassed in finding 'doers' and givers. It was one of her immense strengths.

Dr John Rourke, President of the Botanical Society of South Africa, Former Head of Compton Herbarium, SANBI, Friends of Kirstenbosch Newsletter, *Botanical Society, March 2005*

Majestic Table Mountain, flanked by Devil's Peak and Lion's Head, a 'hotspot' for fynbos plant species, viewed from Bloubergstrand, with *Didelta carnosa* in the foreground

CONTENTS

Below: A fynbos community in the Cape of Good Hope Nature Reserve, with the golden conebush *Leucadendron laureolum* in the foreground (see page 31)

Opposite: *Mimetes cucullatus* (see page 35)

1. AN INTRODUCTION TO THE FYNBOS

A great deal of interest was aroused in 1990 when it was reported for the first time that smoke, derived from burning plant material, broke dormancy and stimulated germination of the seeds of a rare and threatened species of fynbos called false heath, *Audouinia capitata*, from the Bruniaceae family (see page 3). Researchers at Kirstenbosch followed up the initial report and produced experimental results to indicate that seeds of many other fynbos species showed a similar response to smoke. This was an important new ecological discovery and prompted questions from botanists around the world as to 'What exactly is fynbos?' 'What is the significance of smoke in plant regeneration in the fynbos?' 'Does smoke have an effect on seeds of plants from other fire-prone communities in different parts of the world?' and 'How important are the effects of heat and the other factors associated with fire in promoting seed germination and plant regeneration?'

The Cape Floristic Region

Fynbos is a unique type of vegetation, which is dominant in the Cape Floristic Region (CFR) in the south-western Cape, at the southern tip of Africa. It occurs typically on sandstone soils. A second distinctive vegetation type in the CFR is

Angus Wilson

Opposite: Fire engulfs a stand of mountain fynbos

Opposite below: A charred fynbos landscape emerges as thick smoke begins to dissipate

Below: The false heath, *Audouinia capitata* in habitat, Cape of Good Hope Nature Reserve (left), and burnt, open cones of the sugarbush, *Protea repens* following seed dispersal (right)

the renosterveld, which is usually restricted to richer fine-grained soils. Renosterveld shares few species with fynbos, although the two vegetation types often grow adjacent to one another.

The CFR covers an area of 90 000 km^2 (35 000 sq. miles), which is less than 4% of the area of South Africa, yet it contains 9 000 plant species. It is one of the world's richest regions in terms of its biodiversity and over two-thirds of the plant species and 7 of the plant families are endemics, that is, they are not found growing naturally anywhere else in the world. Fynbos, which is a community of small shrubs, evergreen and herbaceous plants and bulbs, is exceptionally rich in species. It is perhaps best known as the home of the South African proteas (sugarbushes, pincushions and conebushes) and ericas (Cape heaths). It is also typified by other very characteristic families such as the Restionaceae (Cape reeds or Cape grasses, which are evergreen rush-like plants) and the Bruniaceae (branching, fine-leaved, heath-like shrubs with characteristic flower heads) and the Cape spring and winter flowering bulbs.

Many of the species from these fynbos families are well known in parks and gardens around the world and as important floricultural crops. Of concern to plant conservationists is the fact that at least 1 491 of the 9 000 species are also Red Data Book species and are in need of conservation. Propagation of some fynbos

Hannes de Lange

plants from seed is difficult, as seeds of many species are dormant when they are shed and they then require very specific environmental 'messages' or cues before they will germinate.

Fire and the fynbos environment

The fynbos occurs in areas with a Mediterranean climate (winter rainfall) and the environment is characterized by a number of stress factors such as summer drought, low soil fertility and periodic fires. The fires have a frequency of 4 to 40 years and are a natural phenomenon in fynbos. Fynbos fire regimes are similar to those of *chaparral* (California), *kwongan* (Australia), *maquis* and non-Mediterranean heathlands, with fires occurring mainly in the summer. Fynbos species are variously adapted to recurrent fire cycles and characteristically experience intense recruitment immediately after fires with little or no recruitment between fires.

Seeds of many species are adapted to germinate in response to one or more of the cues provided by fire. Heat from flames may fracture the impermeable seed coat of hard-seeded species resulting in the coats becoming permeable to water (e.g. Fabaceae or legume family). Dry heat may also break dormancy by providing a heat-pulse that stimulates the seed embryo directly and results in germination (e.g. Restionaceae). Dry heat has also been reported to break seed dormancy of some South African leucospermums (pincushions, family Proteaceae) by complete desiccation of their oxygen-impermeable seed coats. When rain falls the dry coats, which are permeable to water, split suddenly and the embryo then obtains sufficient oxygen for germination. In addition to heat, fires provide chemical messages or cues, such as the gases of ethylene and ammonia that stimulate germination in some species of *Erica*. In addition to the more obvious cues provided by heat, smoke from fynbos fires provides a chemical message that is responsible for stimulating the germination of seed of many fynbos species. The major constituent of this chemical cue was identified by Australian and South African researchers as a butenolide compound as recently as 2004.

To date, 283 species from 39 families have been tested and 161 of these have given a positive response. (See Appendix Table 1 on page 186). Amongst those responding are the horticulturally important species of the Cape reeds (Restionaceae), everlastings (Asteraceae) and brunias (Bruniaceae). Fire may also have an indirect effect on germination by causing changes in the day and night soil temperatures in the immediate post-fire environment. Fire thus provides the major cues for germination in the wild and these cues have to be simulated when attempting to germinate wildflower seed in the laboratory and nursery.

For a complete list of references used in this guide and recommended further reading, see pages 182–184.

Opposite above: Extent of the Cape Floristic Region (CFR) – shaded area

Opposite below: The black-bearded protea, *Protea lepidocarpodendron* regenerates from seeds dispersed following wild fires (see page 188)

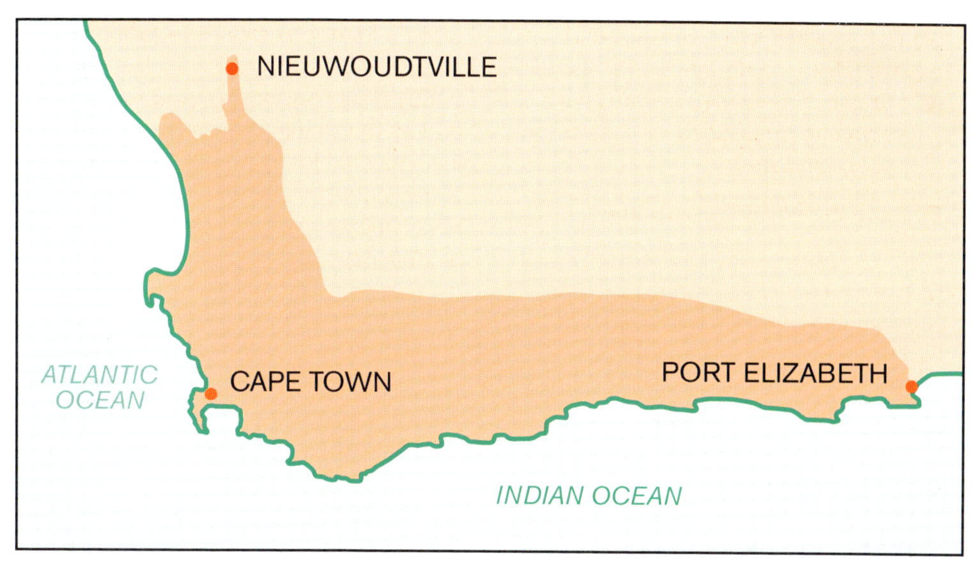
NIEUWOUDTVILLE
ATLANTIC OCEAN
CAPE TOWN
PORT ELIZABETH
INDIAN OCEAN

Below: The king protea, *Protea cynaroides* is one of the easiest and most adaptable proteas in cultivation, and the national flower of South Africa (see page 18)

Right: *Leucadendron tinctum* (see page 33)

Gwynneth Glass

2. GROWING FYNBOS PROTEAS

(With contributions from Gert Brits, Deon Kotze and Philip Botha)

Propagation from seed

Seeds of this family can be divided into two distinct types. One type is rounded, relatively hard and nut-like. Under natural conditions these seeds are shed when mature, collected by ants, carried underground and end up being stored in the soil. In the second seed type, the seed is winged or hairy, often feathered, and relatively soft. This seed type is shown by the so-called serotinous species, i.e. species in which the seeds are stored in the living plant canopy. Serotinous genera comprise *Protea, Aulax* and most of the leucadendrons and make up 37% of the Cape Proteaceae; most of the remainder are nut-like. Nut-like and serotinous seeds show different germination syndromes or patterns.

Proteaceae with nut-like seeds (e.g. *Leucospermum*)

In nut-like seeds the first indication of germination is shown by a crack in the brittle seed coat, caused by the expansion of the cotyledons (first leaves). The young root then emerges through the crack. Under natural conditions, these seeds germinate *en masse* during the

Above: The sugarbush *Protea repens* at the Cape of Good Hope Nature Reserve (see page 21)

Below: *Protea compacta* in the Protea Garden at Kirstenbosch (see page 18)

first winter after a fire. Seed dormancy is broken when the seeds are subjected to moderately low temperatures. This low temperature requirement is not a stratification requirement (see glossary on page 185) but a natural mechanism to promote germination during the favourable cool, moist western Cape winter period. Fluctuating day and night temperatures are also required for maximum germination. For example, a night temperature between 4°C–10°C followed by a day temperature of 20°C–24°C will give optimum germination.

The hard seed coats of the nut-like seeds tend to restrict the supply of oxygen to the seed embryo and this slows down germination. *Leucospermum* seed germination can be improved by pre-soaking seeds in a 1% solution of hydrogen peroxide (H_2O_2) for 24 hours (hydrogen peroxide solution can be obtained from your local pharmacy).

Leucospermum seeds also have a relatively soft, gelatinous outer seed coat, which cuts down the oxygen supply to the embryo. This coat should be removed before sowing. This can be done by soaking seeds in 1% H_20_2 (hydrogen peroxide) or water for 24 hours and then rubbing them between the fingers to remove the softened outer layer. In some species of *Leucadendron* germination can be improved by mechanically scarifying the seeds lightly with sandpaper. This treatment improves oxygen supply to the embryo.

In some species of *Leucadendron* smoke treatment improves germination significantly. A smoke treatment combined with a scarification treatment gives even better germination results. Some growers combine H_2O_2 (hydrogen peroxide) treatment with commercial smoke primer paper by washing each smoke primer paper with 50ml 1% H_2O_2 and soaking seeds in this solution for 24 hours.

Seeds tend to lose viability with age. If seeds are of uncertain age or viability, germination can be stimulated to some extent by pre-soaking seeds in a solution containing germination stimulating substances such as gibberellin GA_3 or GA_4 and GA_7 or ethrel. (These are available commercially from suppliers of scientific and agricultural chemicals). Seeds should also be given a light dusting with a fungicide dressing to prevent post-emergence seedling infection.

Serotinous Proteaceae (e.g. *Protea*)

In serotinous species the much softer seed coat does not crack but splits as the young root emerges. Serotiny is an adaptive response to fires that occur in regular cycles of 4 to 40 years in fynbos vegetation. Fires occur in the late summer or autumn. Mature seeds are held in the living canopy until the canopy is burnt. When the seed heads are burnt or dried by the fire they open and the seeds are shed (see page 3). Seeds have a low temperature requirement for germination. They require temperatures of between 1°C–11°C and this allows the avoidance of drought in the habitat by synchronizing germination with the first wet winter season following dispersal.

Seeds are very sensitive to excess water and it is essential that they be sown in a well-aerated, well-drained sandy soil in order to avoid waterlogging. Another alternative is to sow seeds between wet hessian sacking material and to plant seedlings out when the young roots emerge. Seeds should be given a light dusting with fungicide dressing (e.g. Kaptan) to prevent post-emergence seedling infection.

Vegetative propagation

Preparation and rooting of cuttings

Proteas can also be propagated vegetatively by means of cuttings. All hybrid cultivars are propagated in this way in order to maintain the unique characteristics of each cultivar. The most suitable cuttings are firm shoots that have just completed their growth period. Stock plants should be healthy and disease free and should not yet be flowering. Erect terminal or upright-growing lateral shoots (200 to 250mm long) are preferable to horizontal side-shoots. Leaves on the basal third of the cutting should be cut off and it is necessary to ensure that some leaves with buds remain, and that the cutting has been cut just above a

Percy Sargeant

Above: The sugarbush, *Protea repens* is tolerant of a wide range of soils, from heavy clay to deep sand (see page 21)

potential sprouting bud. The best time to take cuttings is from November to the end of April (early summer to mid autumn). All cuttings need to be well rooted and ready for planting out before August (early spring).

The base of each cutting should be dipped in a solution of rooting hormone for 5 to 10 seconds. Indolebutyric acid (IBA) powder at 4g/l, dissolved in 50% ethyl alcohol can be used for this purpose. Commercially available rooting agents, such as 'Seradix 2' powders or the liquid 'Dip and Grow' can also be used.

Each cutting should be rooted in a perforated transparent plastic sleeve filled with a suitable rooting medium. Alternatively, cuttings can be placed in multi-trays. The rooting medium that gives the best results is a well-aerated mixture consisting of equal quantities of peat or milled pine bark and polystyrene beads. A mixture of equal quantities of peat or milled pine bark and coarse sand is also satisfactory. Rooting is then done under standard mist-bed conditions, with bottom heat (22°C–25°C) provided. Free air circulation around the containers is necessary to minimise harmful fungal growth. A regular spraying programme using a general fungicide may be necessary to control fungi. Any infected or wilted cuttings should be removed and destroyed. Rooting time varies from 8 to 16 weeks. Rooted cuttings should be hardened off under 50% shade cloth for 3 to 4 weeks before being placed in full sunlight. Cuttings rooted in perforated plastic sleeves are ready for planting out once the new roots have developed strongly. Alternatively, they can be established in bags or pots before being planted out. Cuttings rooted in multi-trays should first be established in bags or pots before being planted out.

Grafting

Grafting in proteas is the process of inserting a shoot (scion) into a rootstock, and is advantageous in the cultivation of certain species having rootstocks sensitive to clay, lime and wet soils, and root diseases, allowing them to be grown on rootstocks of species tolerant of these conditions. The choice of rootstock is important as certain species are only compatible with certain rootstocks. Suitable rootstock material includes *Protea eximia* (drought tolerant, compatible with *P. grandiceps*), *Protea obtusifolia* (lime tolerant, compatible with many species), *Protea compacta* (adapted to wet soils, compatible with *P. aristata, P. grandiceps, P. magnifica* and *P. pudens*), and *Leucospermum conocarpodendron* (tolerant of clay soils, compatible with *Mimetes, Orothamnus* and *Serruria*).

Grafting is best done in early autumn from new growth that has hardened-off. Scion and rootstock material used must be disease-free and it is essential that working surfaces be disinfected with sodium hypochlorite (Jik). Hands should be kept clean and secateurs and scalpels regularly disinfected with 50% alcohol. Wedge grafting is the most successful technique for proteas and is best performed on rooted cuttings. Once established, the leaves and buds in the upper 5 cm of the rooted cutting are removed, as well as the upper 1 cm of the cutting. An incision of 2 cm is then made down the centre with a sharp scalpel. Scion material is prepared by cutting shoots 4 cm long that are the same thickness as the rooted cutting. Two buds and two leaves should be present in the uppermost portion of the scion, and the leaves should be cut back to 0.5 cm long. All other leaves and buds are removed. The base of the scion is then cut to a wedge shape such that it fits neatly into the cut in the rootstock, ensuring that the cambium layers of the scion and rootstock are joined, then tightly sealed with grafting tape. The grafted plants are placed in 50% shade, and after new growth on the scion has reached 5 cm, the grafting tape can be removed and they can be planted out into the garden.

Cultivation of fynbos Proteaceae

Soil conditions

The vast majority of South African Proteaceae occur in the south-western Cape, hugging the coast and mountainous regions eastwards to Port Elizabeth and spreading up the west coast to Vanrhynsdorp. The soils are variable, generally poor, with a predominance of soils derived from Table Mountain Sandstone, particularly in the mountainous regions. Species of Proteaceae also grow on Bokkeveld shale, which has a high content of clay. Along the coastal regions, proteas are often found growing in pure sand. The pH values of the soils are usually on the acid side, although there are instances of proteas growing in areas with alkaline soils with pH values as high as 8. Probably one of the most important requirements for the successful cultivation of Proteaceae is a well-drained soil that is also well aerated. Under natural conditions proteas are mostly found growing on mountain slopes that provide such well-drained sites.

Climatic conditions

In their natural habitats proteas are found growing in areas that show a considerable variation in temperatures. For example, a maximum of 32°C is not uncommon during the summer, particularly in the Sandveld and Cederberg. Temperatures are lower in the mountain ranges where the effects of prevailing winds, mists and cloud have a cooling effect. The minimum temperatures in these mountainous areas may fall below 0°C, but not usually for long periods of time. Snow falls regularly on the Cape mountains during winter.

Proteaceae occur in regions where the rainfall varies from as low as 180 mm to 2500 mm per annum. This, however, is rather deceptive. In their natural habitat many species occur in depressions, gullies, valleys and on south-facing slopes where the plants utilize underground moisture through seepage accumulated during the winter months. A good example of this is *Protea cynaroides*, which is always found growing in areas with abundant underground seepage.

Members of the protea family are essentially social plants, although there are some exceptions. Many of the species growing in their natural habitat occur in close proximity to one another, forming close-knit communities. The individual plants protect one another from prevailing winds. A dense cover is established that prevents compaction, keeps the soil cool and reduces the rate of evaporation.

Plants generally are adaptable and the Proteaceae is no exception. With an understanding and appreciation of the basic growing requirements of these plants one is assured of a reasonable chance of success and a great deal of pleasure. In the parts of the United Kingdom and Ireland where there are milder winters some Proteaceae have been grown successfully in the open, however in most areas the winters are too cold and proteas have to be grown under greenhouse conditions.

General sowing directions

Seeds are best sown in late autumn and may be sown in trays or in the open ground in mild areas. Seed trays should be well prepared with plenty of drainage holes covered with a layer of roughage to prevent blockage. It is important that the soil mixture be well drained. A suggested soil mixture is two parts coarse sand, one part leaf mould and one part loam.

The positioning of trays in the greenhouse or in the open is important. A sunny situation with good air circulation helps a great deal in preventing the seeds from becoming waterlogged in the trays and also helps minimize fungal infection.

Seeds should be sown evenly, then firmed down and covered with dry sand and watered thoroughly. Apply a further layer of sand to cover any exposed seed. A suitable depth for sowing is one and a half times the size of the seed.

The optimum temperatures for germination are night temperatures of 5°C–8°C alternating with day temperatures of 15°C–20°C. The germination period varies from 1 to 3 months, depending on the species. The cotyledons are the first to appear, followed by the true leaves. At this stage the young seedlings are ready for pricking out into individual containers.

Plastic bags, 500 ml in size, are most suitable. A soil mixture, similar to that prepared for the seed trays, is an ideal growing medium.

After seedlings have been pricked out, they should be watered thoroughly and then placed in containers in full sun. If conditions are particularly hot and dry, a period of hardening off under light shade cloth may be required. Within 6 months the young plants will have grown to a height of about 200 mm. The roots will show signs of growth through the bottom of the containers and the plants will therefore be ready to be planted out into the open ground or into permanent containers.

When choosing a site for growing Proteaceae in the open, ensure that you have an adequate supply of water until plants are established, a well-drained acid soil, a sunny aspect and good air circulation. Before planting, the chosen site should be cleared of all growth, and individual holes, 250 mm deep and square, prepared for each plant. Manure should be avoided. Well-matured compost mixed with soil provides a suitable growing medium for the young plants. Place the soil mix around the plants at planting. The application of organic fertilizers, derived from seaweed, will encourage the development of the sensitive proteoid roots. Artificial fertilizers must be avoided, especially those containing phosphorous (P) and potassium (K) as these are absorbed in excessive quantity by the proteoid roots, resulting in collapse and death of plants.

Planting distance

The recommended planting distance is 0.65 m for all species that attain a maximum height of 2 m. All species exceeding 2 m are planted at a distance of 1 m. During the first 2 years the young plants must be watered regularly. By mulching the area with compost or wood chips to a depth of 50 mm, weed development is kept to a minimum. Other advantages are that the soil is kept moist, soil temperatures are kept down and mulching is a form of feeding.

Pruning

Pruning is essential in maintaining the desired shape and extending the life of proteaceous plants. It should begin at the seedling/cutting stage and continue throughout the plant's life. Two types

Opposite: The green sugarbush *Protea coronata* (see page 188)

Below: *Mimetes cucullatus* in habitat, Elgin (see page 35)

of pruning are used: thinning-out and heading-back. Thinning-out is the removal of excess, diseased or dead branches at their base. Heading-back is the removal of branches at any point above the base, to encourage resprouting. In heading-back, branches of most species must always be cut at a point above where healthy leaves occur in order for strong new growth to develop from the leaf axils; exceptions are species with lignotubers like *Protea cynaroides* and *Protea speciosa*, in which branches are severely headed-back to the base of the plant, to encourage resprouting. Mature proteaceous plants are best headed-back directly after the flowering period, just before active vegetative growth begins.

Seedlings of slow-growing, well branched species like *Protea grandiceps*, *Protea magnifica* and most *Leucadendron* and *Leucospermum* species can be thinned-out to the strongest 3–5 shoots after one year, then headed-back in the latter two genera, but not in *Protea*. In poorly branching species like *Protea compacta* and *Serruria florida*, heading-back in seedlings is done in the first year during the active growing season. Protea plants grown from cuttings producing a single growth shoot should be headed-back by removing the tip once it has reached 15–20 cm long, while cuttings producing multiple growth shoots should be left alone. In *Leucadendron* and *Leucospermum*, cuttings are best thinned-out in early spring to about 5 strong shoots, then headed-back. If flowers develop on plants grown from cuttings shortly after planting, they should be removed immediately to allow vegetative growth to take place. Large established plants are generally thinned-out or headed-back about once every two years, if necessary. Large pruning wounds should be treated with a wound sealant.

Greenhouse cultivation

Successful cultivation of proteas under greenhouse conditions requires the simulation of a Mediterranean-type climate. This means that for winter conditions temperatures should not drop below 0°C and average minimum winter temperatures should be 5°C–8°C and average maximum winter temperatures 15°C–20°C. For summer conditions minimum temperatures of 12°C–15°C and maxima of 25°C–30°C are ideal. Humidity should be kept low and there should be adequate movement of cool air. Hot and steamy conditions should be avoided at all costs.

An excellent manual on the cultivation of proteas in greenhouses by the English gardener Joseph Knight appeared in 1809, and his advice is still relevant today (visit http://protea.worldonline.co.za/growknight.htm)

A practical approach is to grow proteas in containers and grow them out of doors during the warmer part of the year and then bring them into the greenhouse for protection during the winter.

Pest and disease control

There are obvious advantages to cultivating proteas in their natural habitat in the south-western Cape in South Africa, where the climate and soil conditions are generally favourable. However, it is also a fact that pests and diseases that have evolved together with African Proteaceae are often a bigger problem in their land of origin than in other countries. These pests and diseases do, however, have the potential to become a problem not only in the protea production areas in South Africa, but also where proteas are grown in other countries. It is therefore important to implement the necessary control measures to ensure healthy, disease-free plants.

Control measures start with good sanitation practices in the nursery, orchard and greenhouse. The incidence of fungal invasion above and below the ground in Proteaceae is always related to temperature and humidity conditions. Where temperatures are above 18°C and this is accompanied by a relative humidity of 75% or more, fungal invasion can be expected in most varieties. Proteaceae plants, especially the grey, blue-grey and hairy-leafed varieties, cannot tolerate such conditions and can be badly damaged or killed by fungal invasion. Insect infestation of plants is also largely governed by the climate. Insects either thrive or their numbers remain static depending on the temperature and moisture conditions prevailing. There are a number of ways to combat fungi and insects by paying attention to air movement, light levels and selecting the correct variety or variant for a particular location. Natural and biological measures are often sufficient to enable Proteaceae to be grown satisfactorily in a garden but when they are being grown commercially it is usually necessary to use chemical controls to ensure that they are kept clear of any infestation.

Contact and systemic chemicals are the two basic types used to control fungi and insect pests. Contact sprays work by forming a total cover of chemical that envelops the plant and is effective until it becomes oxidized or weathers off. The systemic types work by entering the plant's sap system. Plants should never be sprayed with any chemicals when they are under stress from dehydration or from any other causes. In addition they should not be sprayed in full sun if the temperature is above 25°C. The safest and best cover is obtained when sprays are applied in the morning just as the foliage is drying off from overnight watering or dew.

Phytophthora root rot is the most important root disease of cut-flower proteas. The pathogen *Phytophthora cinnamomi* is present in many soils throughout the world and it affects Proteaceae plants wherever they grow. Spores are waterborne and multiply very rapidly in wet soils at temperatures above 18°C (e.g. during heavy summer rains). The most important symptom of root rot is the wilting of the entire plant from the bottom leaves upwards. Removal of the external bark exposes a dark patch of rotten bark extending from the crown into the root. The use of appropriate systemic fungicides, such as Aliette, is advised, but applications are regarded as a preventative measure rather than a cure for *Phytophthora*. Stem cankers are sometimes misdiagnosed as *Phytophthora* (see page 16).

Aerial fungi that invade Proteaceae include *Botrytis* (blight; grey powdery mildew develops at the soft growth tip or edges of the young leaves), *Pestalotia* (brown leaf lesions), *Drechslera* (blight; brown leaf lesions), 'silver leaf' fungi and *Septoria* (orange spots with dark edge on leaves). The best control for fungal diseases is to provide good air movement around the plants. However, during high-risk periods this will have to be backed up by the application of chemicals. It is better to anticipate problems and apply fungicides as a preventative measure rather than to have to eliminate them once they have become established. The application of contact and systemic fungicides will control the fungi provided they are applied as a regular maintenance treatment during high-risk periods.

Stem cankers are caused by fungi and are sunken patches that are darker in colour than surrounding tissues. They result in death of branches and eventually the whole plant may die. The fungi enter through wounds such as those made by pruning, picking and insects, and is prevalent in moist weather. Its incidence is reduced by sterilizing pruning tools and controlling insect attack.

'Water-soaked spot and leaf collapse' also occurs but it is not clear whether this is a physical disorder which is then invaded by fungi or whether it is an unidentifiable fungus which causes the condition. The problem is most prevalent in leucadendrons but occasionally shows in proteas and leucospermums. It can occur in both field and nursery situations and first shows as a water-soaked patch on leaves. Treatment with fungicides at three-day intervals appears to restrict its progress.

Soil-borne pests like nematodes are the most serious pest problem in Proteaceae, particularly in warmer climates. There are soil treatments that can be carried out to control their presence but once they are established in adult plants there is little that can be done to eradicate them.

Other soil-borne pests that feed on root tissue may cause problems, especially on plants that are less than three years old. Some caterpillars feed on the young roots of plants to a depth of 30 cm.

Crawling or flying insects permanently resident on plants or found within the nursery or greenhouse are another source of damage. There are three basic types. Firstly, the chewing and sucking ones such as leaf rollers, loop caterpillars, weevils and aphids; secondly, the mites and thrips and thirdly, scales. Each of these may become a problem depending on the season of the year. Low population levels are present at all times on either Proteaceae or other adjacent host plants.

In the case of chewing and sucking insects the eggs are laid on the underside of the leaf, in the flower head or, in the case of the leaf roller, within the terminal tip. When they hatch the caterpillars feed on the foliage. Control with the application of insecticides, usually of a systemic nature. Mites and thrips can become a serious problem on many varieties of *Protea*. Populations normally reach a peak by late autumn each year. Severe infestations are difficult to control and it is usually necessary to spray several times with a mite-eradicating preparation such as Oleum.

There are several different types of scale, all of which live on the underside of the leaves or occasionally on the stems, but always out of direct sunlight. Like thrips and mites, they reach peak population levels in the late autumn and will persist through the winter, causing dehydration of the leaves. It has been found that reasonable control can be achieved safely by using an insecticide in combination with a good wetting agent and applied as a spray on its own and not in combination with a fungicide.

Protea neriifolia 'Red Robe' (foreground) makes an excellent cut-flower (see page 20)

Ten of the Best Fynbos Proteas

(see also Appendix 2, page 188)

Protea compacta

Common name: Botrivier protea

Height: up to 3.5 m

Flowering time: autumn to spring

An erect, sparsely branched, lanky shrub, with long cup-shaped flower heads, bright pink, occasionally white. This striking species has a very long flowering season and is a prolific bloomer, commencing flowering in its third year from seed. It requires acid soil in full sun and is best grown in groups to provide support for the long branches and prevent plants from falling over in strong winds. Young plants should be staked and pinched back. Highly recommended for large fynbos gardens, and an excellent cut-flower. Sow seeds in late autumn.

Protea cynaroides

Common name: king protea

Height: 0.3–2 m

Flowering time: all year round

This is one of the most outstanding proteas. It has very distinctive leaves and huge flower heads up to 0.3 m in diameter. Flowering usually takes place in the fourth or fifth year from seed. Plants have a swollen stem base just below the soil surface called a lignotuber from which shoots will sprout if the stems are cut back. It is a very variable species and there are a considerable number of local forms with variations in flower colour and in leaf characteristics. Under natural conditions it always grows on acid sandy soil derived from Table Mountain Sandstone. Although it thrives in well-drained sites, it is a water-loving species and grows naturally in damp areas. It requires full sun and is a long lasting cut-flower. One of the easiest and most adaptable proteas in cultivation, and is suitable for growing under glass. It is mildly tolerant of frost down to -3°C. Sow seeds in late autumn. The cultivar *Protea cynaroides* 'Red Rex' is particularly attractive and easily grown.

Protea eximia

Common names: broad-leafed sugarbush, breë-blaarsuikerbos

Height: 2–5 m

Flowering time: summer

An erect, sturdy shrub, sparsely branched, with broad greyish-green to purplish-green foliage. The trunk may reach up to 300 mm in diameter and the flower heads are large with brilliant red bracts and a purplish-black centre. It is easily cultivated, requiring full sun and acid, well-drained soil and is suited to medium-sized or large gardens. Light pruning after flowering encourages bushy growth. Highly attractive to nectar-feeding birds and large metallic-green scarab beetles. Sow seeds in late autumn. The cultivar *Protea eximia* 'Fiery Duchess' (above) is a strong grower, drought tolerant, an excellent cut-flower and is mildly tolerant of frost down to -3°C.

Protea grandiceps

Common name: red sugarbush

Height: up to 2 m

Flowering time: spring to summer

This is an attractive compact, tidy, rounded shrub. It has broad oval blue-green leaves and numerous bright coral-pink flower heads with a white or purple fringe to each involucral bract. Although slow-growing, it is a very decorative garden shrub and a long lasting cut-flower. It requires well-drained acid soil in full sun and tolerates frost down to -5°C. Sow seeds in late autumn.

Protea magnifica

Common names: queen protea, baardsuikerbos

Height: 0.5–2.5 m

Flowering time: midwinter to midsummer

A robust, sprawling shrub with striking blue-green or greyish, strap-like leaves that are hairy when young but become hairless with age. It has magnificent, large flower heads, with heavily bearded white to deep carmine involucral bracts surrounding the hairy flowers clustered at the centre. It is a slow-growing species requiring very well-drained acid soil in full sun, and flowers in its third or fourth year from seed, under ideal conditions. It is an excellent cut flower and tolerates frost down to -5°C. Sow seeds in late autumn.

Protea neriifolia

Common names: oleander leaf protea, blue sugarbush

Height: 3 m, sometimes to 5 m

Flowering time: differs between local variants and may be from late summer through winter to late spring

Although in nature *Protea neriifolia* grows in a fairly moist and moderate climate and in sandy acid soils derived from Table Mountain Sandstone, in cultivation it has a wider tolerance of soils than most other proteas, and is consequently much sought after by growers. The floral bracts range from pure white to pink to a deep rose or wine colour and their beards also vary in colour. Flowering commences in the fourth year from seed. It is probably the most satisfactory protea for the average garden, not only as an ornamental shrub, but also as a generous provider of cut-flowers. It is also highly attractive to nectar-feeding birds. Sow seeds in late autumn. The cultivar *Protea neriifolia* 'Red Robe' (see page 17) is particularly attractive and tolerates mild frost down to -3°C.

Protea pudens

Common name: bashful sugarbush

Height: up to 0.4 m

Flowering time: midwinter to early spring

A beautiful dwarf protea forming a prostrate shrub with a spread of up to 1 m. It produces a single main stem with trailing horizontal branches and the numerous erect linear leaves usually have hairy margins. The bell-shaped flower heads have deep brownish-pink involucral bracts and a prominent black woolly cone in the centre. It is an endangered species in its native habitat in the southern Cape. It responds well under cultivation and is an ideal subject for rock garden pockets in acid sandy-clay soil in full sun, or for sprawling over low retaining walls or planted in large, deep containers. Sow seeds in late autumn.

Protea repens

Common name: sugarbush

Height: 1–2 m

Flowering time: all year round, mainly autumn to spring in the Western Cape and spring to late summer in the Eastern Cape

In its natural habitat, this is one of the most widely distributed of all Cape proteas. They have attractive flower heads with involucral bracts that are sticky and shiny. There are a number of colour variants, ranging from white to deep crimson or, alternately flowers are white with pink tips. Plants flower in the third year from seed and time of flowering varies depending on the variant. They are waterwise and tolerant of a wide range of soils, including less acidic soils, from heavy clay to deep white sand. For the average gardener *P. repens* is one of the easiest, most adaptable and reliable proteas in cultivation and is also suitable for growing under glass. It is highly attractive to nectar-feeding birds. Sow seeds in late autumn. There are several outstanding cultivars, of which *P. repens* 'Rubens' (above) is one of the most desirable, and tolerates mild frost down to -3°C.

Protea scolymocephala

Common names: witskollie, thistle sugarbush

Height: 0.3–1.3 m

Flowering time: late winter to late spring

A much-branched, low-growing shrub with attractively flared, yellowish-green flower heads and narrow, greenish-grey foliage. Ideally suited to acid sandy soils in windy coastal gardens, or to large, deep containers. It is a fast grower, waterwise, and a fairly long-lived species, well suited to rock garden pockets. Light pruning of the branches keeps the plants looking neat and prevents them from becoming too lanky. As a cut-flower, its small, long-lasting blooms are excellent for small arrangements. Sow seeds in late autumn.

Protea speciosa

Common names: brown-bearded protea, baardsuikerbos

Height: 1–1.2 m

Flowering time: late winter to late spring

A slow-growing, but long-lived small shrub having attractive satin-pink inflorescences with the involucral bracts thickly bearded with brown, produced on stout hairy stems. Its greenish-grey leaves are very leathery and the plant produces sprouts from the base if the stems are cut back. It requires acid, well-drained soil in full sun and tolerates mild frost down to -3°C. It is an excellent cut-flower and highly recommended for small fynbos gardens. Sow seeds in late autumn.

Ten of the Best Fynbos Pincushions

(see also Appendix 2, page 188)

Leucospermum bolusii

Common names: Gordon's Bay pincushion, witluisiesbos

Height: up to 1.5 m

Flowering time: late spring

An erect to spreading, neat rounded shrub with many long branches arising from a single main stem. An abundance of small, fragrant creamy-white flower heads are produced. It is a decorative, long-lived shrub requiring acid soil and full sun conditions, ideally suited to the small fynbos garden, and is a good cut-flower. Sow seeds in late autumn.

Leucospermum cordifolium

Common name: pincushion

Height: 1.5 m

Flowering time: spring to summer

This is one of the best-known members of the genus and probably the first to be called 'pincushion'. This species is widely cultivated both as a garden shrub and for cut-flowers and is suitable for growing under glass. It forms a rounded, spreading shrub up to 2 m across and plants have a single main stem and horizontally radiating branches. Highly ornamental blooms decorate the shrubs for up to 6 to 8 weeks during spring. Inflorescences range from yellow to pink to deep orange or crimson and there are numerous excellent cultivars. Highly prized by nectar-feeding birds. Plants require acid soil, full sun, are fast growing and flower within 3 years from seed. Best sown in late autumn.

Leucospermum erubescens

Common names: Oudtshoorn pincushion, oranjevlamspeldekussing

Height: up to 2 m

Flowering time: spring to summer

An erect, strong-growing shrub with oblong shaped leaves, producing numerous large heads in clusters of 4–8 at the top of flowering branches. During the flowering season the heads change from pale yellow to brilliant crimson as they mature. Recommended for arid fynbos gardens as it tolerates low winter rainfall conditions but is easily grown in high winter rainfall areas. Requires acid soil and full sun. An excellent cut-flower and highly attractive to nectar-feeding birds. Sow seeds in late autumn.

Leucospermum glabrum

Common name: Outeniqua pincushion

Height: 2–2.5 m

Flowering time: late winter to late spring

A robust, much-branched shrub producing dark green elliptical leaves with 3–6 apical teeth. The very large dark orange flower heads are borne singly at the tips of the branches and have conspicuous red pollen presenters. A floriferous, easily grown species. It is highly recommended for both small and large gardens in full sun, requiring acid soil and is attractive to nectar-feeding birds. An excellent cut-flower. Sow seeds in late autumn.

Leucospermum grandiflorum

Common names: luisiesboom, grey-leaf fountain pincushion

Height: up to 3 m

Flowering time: late winter to midsummer

An erect shrub branching from the base and spreading with age. It has distinctive grey, elliptic leaves with reddish apical teeth, and the large yellow flower heads have conspicuous reddish pollen presenters. It thrives in well-drained granitic soils in full sun and tolerates relatively low winter rainfall. Ideally suited to small gardens and highly attractive to nectar-feeding birds. A good cut-flower. Sow seeds in late autumn.

Leucospermum muirii

Common names: Albertinia pincushion, bloukoolhout

Height: up to 1 m

Flowering time: late winter to early summer

A tidy, rounded shrub with narrow, dark green foliage, reddish stems and showy small yellow flower heads that mature to orange with age. Highly recommended as a low growing shrub for a small fynbos garden or for large, deep containers. It requires acid well-drained soil in full sun and responds well to light pruning after flowering. Sow seeds in late autumn.

Leucospermum oleifolium

Common name: Overberg pincushion

Height: up to 1 m

Flowering time: mainly from spring to early summer

A most attractive compact, rounded shrub branching from the base, with greenish-grey, smooth or hairy foliage. The plants are long-lived and very floriferous and flower heads open yellow, becoming red with age. They require acid sandy soil in full sun and are highly recommended for small fynbos gardens or large, deep containers. Suited to windy sea-side conditions and a good cut-flower. Sow seeds in late autumn.

Leucospermum praecox

Common name: Mossel Bay pincushion

Height: 2–3 m

Flowering time: autumn to spring

An erect rounded shrub up to 4 m across, with the trunk up to 80 mm in diameter. The hairless leaves are oval to wedge-shaped with 5–11 nectar-secreting glandular teeth. The large yellow flower heads turn to orange when mature and it is a very early flowering species with a long flowering period. In its natural habitat, form Albertinia to Mossel Bay, it is locally abundant over large areas where it creates superb floral displays. It is recommended for small to large gardens in acid soil in full sun. A good cut-flower. Sow seeds in late autumn.

Leucospermum reflexum var. *reflexum*

Common names: rocket pincushion, perdekop

Height: up to 4 m

Flowering time: early spring to midsummer

A long-lived, large rounded shrub with silvery-grey foliage and distinctive heads of orange to scarlet flowers with the styles sharply bent backwards and downwards. It is a vigorous, attractive subject suited to large gardens in acid soil in full sun, and is best planted in groups to provide support and prevent falling over in strong winds. A good cut-flower and highly prized by nectar-feeding birds. Sow seeds in late autumn.

Leucospermum reflexum var. *luteum*

Common names: yellow rocket pincushion, geel perdekop

Height: up to 4 m

Flowering time: early spring to midsummer

Like the orange and scarlet forms of *Leucospermum reflexum* var. *reflexum*, this clear yellow variety is an excellent strong-growing subject for large gardens and is most effectively displayed when inter-planted with the orange and scarlet forms (see pages 206–207). It is long-lived, requiring acid soil in full sun. A good cut-flower and attracts nectar-feeding birds. Sow seeds in late autumn.

Leucospermum tottum

Common names: ribbon pincushion, oranjerooi-speldekussing

Height: up to 1.5 m

Flowering time: spring to summer

A very floriferous neat shrub, with a single main stem and horizontally spreading branches. It produces linear-oblong leaves and large solitary flower heads. The perianth is pink or dull carmine to brown and the styles are pale pink. It requires acid well drained soil in full sun and tolerates light frost down to -1°C. Attractive to nectar-feeding birds. Highly recommended for both small and large gardens and a good cut-flower. Sow seeds in late autumn.

Above: *Leucospermum tomentosum* (see page 188)

Below: The silver tree, *Leucadendron argenteum* (see page 29)

Ten of the Best Fynbos Conebushes

(see also Appendix 2, page 188)

Leucadendron argenteum

Common names: silver tree, silwerboom, witteboom

Height: up to 10 m

Flowering time: spring

An attractive fast growing, well proportioned tree with a stout trunk and thick bark. It has large light-reflecting, silver grey leaves, the margins fringed with long hairs. The flowers are dioecious with silvery cream floral bracts. The male inflorescence (above) is 50 mm in diameter and although the female head is only 40 mm across, it matures to a silver cone 90 x 60 mm with bracts densely covered in silvery hairs. The fruits remain in the cone for a year or longer. Each bi-convex fruit or seed is equipped with a 'parachute'-made up of the persisting style and the plumed perianth, which aids dispersal by the wind. The trees are wind tolerant and are attractive garden subjects. They require acid, well-drained soil in full sun and are sensitive to root disturbance. They are highly susceptible to *Phytophthora* root rot. The decorative foliage is useful for flower arrangements. Sow seeds in late autumn.

Leucadendron daphnoides

Common names: Du Toit's Kloof conebush, reusepoeierkwas

Height: 1–1.5 m

Flowering time: midwinter to spring

An erect small shrub, branching from the base of a single main stem. It has lanceolate greenish-grey leaves and those surrounding the showy flower heads turn bright yellow or red during the flowering period. The male flower heads (above) as well those of the females have a citrus-like scent. An easily cultivated and fairly long-lived plant recommended for small gardens with acid, well-drained soil in full sun. It is a good cut-flower. Sow seeds in late autumn.

Percy Sargeant

Leucadendron discolor

Common names: Piketberg conebush, rooitolbos

Height: up to 2 m, rarely up to 4 m

Flowering time: spring

An erect bushy shrub with leathery, greenish-grey oval leaves. The male plants (above) are exceptionally attractive in spring with their yellow involucral leaves surrounding the bright red flower heads. The female plants have somewhat longer leaves and their cone-like flower heads are surrounded by pale green involucral leaves. The plants respond well to cultivation in very well-drained, acid soil in full sun. Both the male and female inflorescences are excellent cut-flowers with long, sturdy stems. Sow seeds in late autumn.

Leucadendron eucalyptifolium

Common names: gum-leaf conebush, grootgeelbos

Height: 3–5 m

Flowering time: winter to spring

A robust shrub or a small tree with a single main stem. It produces linear, eucalyptus-like leaves. The leaves surrounding the scented flower heads are bright yellow and longer. The male flower heads (above) are slightly larger than those of the females, and the latter have a fruity scent. A fast growing, long-lived shrub that flowers in 2 years from seed. It requires acid, well-drained soil in full sun. Recommended for large fynbos gardens as specimen plants or informal hedges, and is useful in flower arrangements. It tolerates mild frost down to -3°C. Sow seeds in late autumn.

Leucadendron gandogeri

Common names: broad-leaf conebush, berggeelbos

Height: 1–1.5 m

Flowering time: spring

An attractive rounded shrub branching from the base with elliptical leaves borne on sturdy reddish stems. The involucral leaves surrounding the female flower heads are larger and clear yellow, and the male flower heads (above) are lemon-scented. It is an easily cultivated, long-lived plant, recommended for small gardens with well-drained, acid soil in full sun, and is an excellent cut-flower. Sow seeds in late autumn.

Leucadendron laureolum

Common name: golden conebush

Height: up to 2 m

Flowering time: winter

This is a shrub that in its natural habitat grows happily in a great range of soils from sea level to 1000 m. Whereas male shrubs are rounded in appearance (above), females are less symmetrical. The involucral leaves, which are bright yellow in colour, partially conceal the flower heads in males and totally conceal those in females. Plants show a moderate rate of growth and flower within 2 years from seed. They produce dazzling floral foliage and are good cut-flowers. They require acid, very well-drained soil in full sun. Sow seeds in late autumn.

Leucadendron macowanii

Common name: acacia-leaf conebush

Height: 2–2.5 m

Flowering time: late autumn to winter

An attractive erect shrub, with willowy yellowish-green foliage. The dense, yellow male flower heads (above) bloom without involucral leaves and the female inflorescences form attractive dark brownish-maroon cones. Although an endangered species in its natural habitat in the southern Cape Peninsula, it is a long-lived, easily grown plant, and likes moist, acid conditions in full sun. Sow seeds in late autumn.

Leucadendron salignum

Common names: common sunshine conebush, geelbos

Height: up to 2 m, but usually much less

Flowering time: variable, from autumn to midsummer

A very variable species with the widest distribution of all the leucadendrons. It is an erect or spreading multi-stemmed shrub with a persistent rootstock. The involucral leaves of male plants are similar to the stem leaves but longer, and red or yellow (see pages 39, 204). In female plants the involucral leaves are usually red or creamy-white (above). It is a good cut-flower and flowers in 2 years from seed. Highly recommended for small fynbos gardens and large, deep containers. It requires acid, well-drained soil in full sun and tolerates frost to -5°C. Sow seeds in late autumn.

Leucadendron sessile

Common name: western sunbush

Height: 1–1.5 m

Flowering time: late winter to early spring

A strong-growing rounded shrub with a single main stem, erect, narrowly elliptical leaves and bright yellow involucral leaves that mature to dark red in summer. The showy male flower heads (above) are much larger than those of the females and are strongly lemon-scented and highly attractive to insects. An easily cultivated, long-lived plant suited to granitic clay soils, and an excellent cut-flower. It requires acid, well-drained soil, moist conditions in winter, and full sun. Sow seeds in late autumn.

Leucadendron tinctum

Common names: spicy conebush, toffie-appel

Height: 1–1.3 m

Flowering time: winter

An attractive bushy spreading shrub with a single main stem. The greenish-grey, oblong-lanceolate leaves are leathery with rounded tips and have no stalks. The involucral leaves are larger and yellowish-green turning to red, and the female cones (above) emit a strong spicy aroma. An easily grown, long-lived leucadendron, ideally suited to mixed fynbos beds, and a favourite garden plant. It requires acid, well-drained soil in full sun. Sow seeds in late autumn.

Ten of the Best Fynbos Pagodabushes, Sceptrebushes and Spiderheadbushes

(see also Appendix 2, page 188)

Mimetes argenteus

Common name: silver pagoda

Height: up to 3.5 m

Flowering time: autumn to winter

An erect, sparsely branched shrub with a single main trunk. The beautiful silvery leaves are clothed with short silky hairs and held at right angles to branches. The orangy-pink flower heads are borne in terminal spikes, the flowers subtended by a carmine to pale mauve leaf held perpendicular to the stem. The plants are slow growing and require well-drained acid soils derived from Table Mountain Sandstone, in full sun or lightly shaded positions. Light pruning of the branches after flowering prevents the plants from becoming too lanky. They are well suited to rock garden pockets and their roots are highly sensitive to soil disturbance. Sow seeds in late autumn.

Mimetes chrysanthus

Common names: golden pagoda, goue stompie

Height: up to 2 m

Flowering time: late summer to winter

An attractive erect, sparsely branched shrub with greenish-grey leaves and a single main trunk. It is very floriferous and produces scented, dense yellow flower heads in terminal spikes and responds well to light pruning after flowering. It is an easily cultivated, long-lived plant highly recommended for rock garden pockets in small gardens or large, deep containers. A good cut-flower and attracts nectar-feeding birds. Sow seeds in late autumn.

Mimetes cucullatus

Common names: common pagoda, common mimetes, red mimetes, rooistompie

Height: 1–2 m

Flowering time: almost all year, but mainly from midwinter to early summer

A spreading shrub with many erect stems produced from the base, densely covered with overlapping yellowish-green leaves. The colourful red and yellow flower heads have prominent creamy-white, silky stamens and the bright red leaves produced at the tops of the flowering branches are especially attractive. The plants require well drained, acid sandy soil in full sun and frequent moisture from autumn to late spring. A long-lived and decorative garden shrub. Recommended for windy sea-side gardens and highly prized by nectar-feeding birds. Sow seeds in late autumn.

Mimetes fimbriifolius

Common names: tree pagoda, maanhaar stompie

Height: up to 4 m

Flowering time: spring to summer

A long-lived tree or large rounded shrub with a single trunk and thick, corky bark. The leaves are fringed with hairs and those produced at the tops of the cylindrical flower heads are an attractive reddish-yellow colour. The plants are slow-growing and require plenty of moisture in winter. It needs acid, well-drained soil in full sun and is a striking species suitable for large gardens as a specimen plant or an informal hedge. Sow seeds in late autumn.

Percy Sargeant

Mimetes hirtus

Common names: marsh pagoda, vleistompie

Height: 1–1.25 m

Flowering time: autumn to early summer, but mainly winter

An erect, branching shrub with a single main trunk. The leaves, like the rest of the plant, are covered with minute hairs, hence the specific name *hirtus*. The flower heads are very prominent and each head is topped with an erect or flattened comb of reduced green or pink leaves. The floral bracts are yellow with red tips and it is a handsome shrub that is much sought after as a cut-flower. Suited to moist, acid conditions and recommended for small fynbos gardens. Highly attractive to nectar-feeding birds. Sow seeds in late autumn.

Paranomus reflexus

Common name: Van Staden's sceptre

Height: 1–2 m

Flowering time: midwinter to early spring

A neat small shrub bearing two distinct leaf types, those in the lower parts of the branches are finely dissected, while those in the upper parts are entire with their margins curved upwards. The creamy-yellow rocket-shaped flower heads are borne singly at the tips of the branches and the individual flowers are distinctly reflexed. It is easily propagated from seed and an attractive and interesting garden plant for both full sun and lightly shaded situations. Sow seeds in late autumn.

Serruria aemula

Common name: strawberry spiderhead

Height: 0.2–0.5 m

Flowering time: winter to spring

A low-growing, spreading shrub with creamy-pink, sweetly scented flower heads. Historically this species occurred in huge stands just outside Cape Town from Rondebosch to Milnerton, and a few populations still exist there along road verges and under powerlines near Milnerton. It is an attractive plant recommended for sunny rock garden pockets and large, deep containers in acid, well-drained soil in full sun. It needs to be re-propagated from seed at regular intervals as it is rather short lived in cultivation. Sow seeds in late autumn.

Serruria elongata

Common name: bottlebrush spiderhead

Height: 0.3–0.5 m

Flowering time: spring

This erect shrub has a lush appearance with erect branches covered with short, heavily dissected leaves. The solitary, fragrant flower heads are oblong and pinkish-buff to purplish-brown and borne on short peduncles at the apex of each long shoot. It makes a decorative rock garden subject but being rather short-lived in cultivation, it needs to be re-sown at regular intervals. It needs acid, well-drained soil in full sun. Sow seeds in late autumn.

Serruria florida

Common names: blushing bride, trots van Franschhoek

Height: 0.8–2 m

Flowering time: winter to spring

This erect shrub is the most beautiful of the serrurias. In nature, where it is not subjected to pruning, the growth form is tall and lanky with clusters of flower heads at the apex of each shoot. The floral bracts of the individual florets of the head are large and showy, accounting for the deep blush within the heads. Despite its rarity in nature, the blushing bride, thanks to cultivation, is well known and valued as a garden plant. Its economic potential is high and it is easily grown from seed and responds well to pruning. It is a quick grower, flowering in 15 months from seed. It is short-lived in cultivation and has to be re-sown at regular intervals. Requires acid, well-drained soil in full sun. Sow seeds in late autumn.

Serruria villosa

Common name: golden spiderhead

Height: 0.3–0.5 m

Flowering time: late winter to midsummer

An erect, neat rounded shrublet found in the southern parts of the Cape Peninsula. Its yellowish-green, heavily dissected leaves are arranged in whorls and have yellow tips, forming a 'halo' effect around the solitary large, yellow flower heads that are produced at the tips of the branches. The flowers are strongly scented and the floral bracts and perianth segments are clothed in thin, silky hairs, hence the specific name. A most attractive species, highly recommended for sunny rock garden pockets and large, deep containers, but needs to be re-sown at regular intervals as it is rather short-lived in cultivation. Requires acid, well-drained soil in full sun. Sow seeds in late autumn.

Leucadendron salignum (male) is highly recommended for small fynbos gardens (see page 32)

Background photograph: *Leucospermum reflexum* var. *reflexum* (orange) and *Leucospermum reflexum* var. *luteum* (yellow) (see page 27)

Protea neriifolia
Leucospermum reflexum var. reflexum
gnifica
Leucadendron laureolum
Paranomus reflexus

Below: Display beds in the Restio Garden at Kirstenbosch

Opposite: The attractive bracts of *Elegia capensis* (see page 50)

3. GROWING FYNBOS RESTIOS

(With contributions from Hanneke Jamieson and Philip Botha)

Propagation of restios

The Restionaceae is a family of evergreen, rush-like plants that is restricted to the Southern Hemisphere. There are about 330 species in Africa (most in the Western Cape), 150 in Australia, 4 in New Zealand and 1 species each in South America and South East Asia. Members of the African Restionaceae are relatively diverse in their seed dispersal mechanisms, which could be implicated in the survival of seeds during or after fires. There are 3 different modes of dispersal. Firstly, wind dispersal of unilocular, indehiscent ovaries that have a persistent perianth, which acts as a wing for the fruit (e.g. *Thamnochortus, Staberoha* and *Calopsis*). The second mode is myrmecochory (dispersal by ants) of fruits containing elaiosomes. The ovary is unilocular and indehiscent, and the ovary wall is heavily lignified (e.g. the 'nut-fruited' restios like *Anthochortus, Mastersiella, Hypodiscus, Willdenowia, Ceratocaryum* and *Cannomois*). These seeds are also serotinous and are retained on the plant until the next season's seed crop is mature. The third mode is the 'basic' condition, which has dehiscent ovaries, with 1 to 3 locules. The seed is released from the ovary after maturation, but it is not known how it is dispersed after release (e.g. *Restio, Ischyrolepis, Askidiosperma, Chondropetalum, Dovea, Nevillea* and *Rhodocoma*).

Germination problem solving

The poor germination achieved with seed of many species has been attributed to the limited seed-set of some species and the difficulty in determining when seeds are ready for harvest. Heat treatment of seeds at 120°C for 3 minutes gave a significant improvement in the germination of seeds of some species. In common with many other fynbos species, restios require alternating high and low diurnal temperatures as a cue for germination, which is stimulated by plant-derived smoke and aqueous smoke extracts.

In 1994 a major germination study of restios was done at Kirstenbosch. Seed of 32 species was screened to obtain an indication of how important the smoke cue was for germination in this family. The results of this study represented the first occasion that comparative germination data for South African species of this family had ever been obtained. Twenty-five of the 32 species tested showed a statistically significant improvement in germination following smoke treatment.

Untreated seeds of 18 of the species responding, showed a high degree of dormancy and only 0.1- 2% germination was obtained. These results suggested that under natural conditions smoke from fynbos fires provided an important cue for triggering seed germination in many species in this family. The 4 species that did not germinate were all myrmecochorus, nut-fruited species, which are thought to require a different or additional cue for germination.

Below: Smoke promotes germination of seeds of *Rhodocoma gigantea*
Left: smoke-treated seeds;
germination = 2 500 seedlings/gm seed
Right: water control;
germination = 10 seedlings/gm seed (see pages 47, 53)

Jeanette Loedolff

Elegia stipularis (female plant) with *Metalasia muricata* at the Cape of Good Hope Nature Reserve (see pages 188, 190 respectively)

A basic guide to seed germination of restios

- Use fresh, mature seed
- Seeds may be pre-soaked in aqueous smoke extract or a commercial smoke seed primer for 24 hours before sowing, or seeds may be smoked once sown in trays. Fill trays with a loam/composted milled pine bark/sand mixture in a proportion of 1:2:2 that will be well drained. Seeds should be covered with a thin layer of milled bark.
- Incubate seeds under alternating night/ day temperatures of 8°C (16h)–23°C (8h) for optimum germination
- Nut-fruited species remain difficult (often seemingly impossible) to germinate in the nursery or laboratory. One suggested dormancy-breaking treatment is to heat fruits to 120°C for 3 minutes prior to pre-soaking in aqueous smoke extract or pre-soaking in a commercial smoke seed primer for 24 hours. In some nut-fruited species a pre-germination storage treatment at 18°C–28°C for several weeks gives improved germination and others will germinate if incubated moist at 18°C–28°C. In general, germination cues for nut-fruited restios require further study, perhaps with more attention to the role of light.

Seed collection

Restios that are wind pollinated flower during spring or late summer, producing seed after a period of 6–11 months. The seed varies from very fine seed like *Chondropetalum tectorum* with about 10 000 seeds per gram to the large nut-like seeds of *Cannomois virgata* with 4 seeds per gram (see illustrations on page 56). Seed collection is fraught with difficulties as there is very little information available for individual species on the season of flowering, on the period required for seed maturation and on the timing of seed drop. Seeds of many species look ripe from the outside but on dissection are found to be green and immature.

Thamnochortus spicigerus stacks used for thatching on the Cape West Coast (see page 54)

Table 1. Germination response of Restionaceae species to plant-derived smoke

Key to responses: *** indicates very marked increase in germination (1000% or more)

** indicates marked increase in germination (100% or more)

* indicates moderate increase in germination (50-100%)

NR indicates no response to smoke; most nut-fruited species; seeds remain dormant; other germination cues probably involved

Seed size: (seeds/gram) Figures in brackets after each species name

Species	Response
Askidiosperma andreaeanum	**
Askidiosperma esterhuyseniae (1 219)	*
Askidiosperma paniculatum (373)	**
Calopsis paniculata (3 333)	*
Cannomois parviflora (27) NR	
Cannomois virgata (4)	**
Ceratocaryum argenteum (9) NR	
Chondropetalum ebracteatum (794)	**
Chondropetalum hookerianum (859)	***
Chondropetalum mucronatum (248)	***
Chondropetalum tectorum (10 550)	***
Dovea macrocarpa (90)	**
Elegia capensis (2 010)	*
Elegia cuspidata (706)	**
Elegia equisetacea (1 250)	**
Elegia fenestrata	**
Elegia filacea (2 200)	**
Elegia grandis (64)	*
Elegia stipularis	*
Hypodiscus neesii (21) NR	
Hypodiscus striatus (22) NR	
Ischyrolepis ocreata (460)	**
Ischyrolepis sieberi (383)	***
Ischyrolepis subverticillata (538)	***
Restio bifarius (364)	***
Restio brachiatus (2 000)	**
Restio dispar (290)	**
Restio festuciformis (4 147)	**
Restio pachystachyus (340)	*
Restio similis (10 000)	**
Restio tetragonus	***
Restio triticeus (1 136)	**
Rhodocoma arida (1 149)	*
Rhodocoma capensis (5 263)	***
Rhodocoma fruticosa (961)	**
Rhodocoma gigantea (725)	***
Staberoha aemula (602)	***
Staberoha cernua (911)	***
Staberoha distachyos	**
Staberoha vaginata (568)	**
Thamnochortus bachmannii (9 080)	***
Thamnochortus cinereus (1 010)	***
Thamnochortus pellucidus (350)	***
Thamnochortus platypteris (408)	**
Thamnochortus punctatus	***
Thamnochortus spicigerus (385)	***
Thamnochortus sporadicus	**
Willdenowia incurvata (8) NR	

Vegetative propagation

Restios can also be propagated vegetatively by division. The stoloniferous species can be most successfully divided just before the crop of new shoots emerges from the ground, generally in early or midwinter. The plants should be divided into fairly large pieces, the roots disturbed as little as possible, and planted out immediately in the open ground or in containers. After transplanting the plants should be kept well watered until the new shoots are growing and the plant has 'taken'. Generally the plants take up to a year to start growing again and do not seem to grow as vigorously as plants raised from seed.

Cultivation of restios

The normal growing season for restios is in the autumn, spring and early summer. The best time to plant restios in both the summer and winter rainfall areas is at the beginning of the rainy season. The plants are planted in holes of 600 mm square and 400-600 mm deep. The soil that is removed from the planting hole should be well mixed with about 2 spades of compost and then replaced in the planting hole. It is recommended that no fertilizer should be added as this might burn the roots. The plants should be planted at the same level as they were in the bags. They must be well watered after planting and after about 6 weeks they should show signs of new growth. Restios, in common with other fynbos species like proteas, do not like to have their roots disturbed and do not like to be planted in small holes in lawns. They are, however, much more robust growers than most fynbos plants and do not seem to be plagued by soil-born fungi or other diseases. The main requirements for successfully growing restios are full sun, a well-drained soil and plenty of air movement. Restios will respond well to regular feeding with low concentrations of nutrients, such as found in granular slow release fertilizers containing nitrogen, phosphate and potassium. They may be fed with standard organic fertilizers such as Seagrow or Kelpak or by sprinkling the surrounding soil with Bounce Back pellets or a small amount of ammonium sulphate during the growing season. Restios respond well to regular watering by producing lush and robust growth. Most species will, however, tolerate periods of drought, as they are adapted to a long dry summer season.

Mulching

Like most other fynbos plants, restios will benefit from a mulch of milled pine bark or rough compost. The mulch keeps the roots cool, reduces water evaporation and inhibits weed growth.

Maintenance

Maintenance consists of the removal of dead stems from the stoloniferous species, like *Elegia capensis* and *Calopsis paniculata,* where the dead stems can be removed every year or two. The dead stems of the tufted species like *Thamnochortus insignis, Chondropetalum tectorum* and *Restio festuciformis* need not be removed. These species produce a new crop of stems in the centre of the plant and push the older stems outward and to the ground. The old, dead stems will be more or less covered by the green stems and need not be removed. In some street plantings, plants have been pruned to remove the old outer stems, leaving the new crop of stems to grow out unhindered from the centre.

Ten of the Best Fynbos Restios

(see also Appendix 2, page 188)

Cannomois virgata

Common names: bergbamboe, besemriet, olifantsriet

Height: up to 4 m

Flowering time: early to midsummer

This is a variable and extremely vigorous, decorative reed-like plant with thick main stems and finely divided side branches. This widespread restio is found growing in the Western and Eastern Cape, and is most commonly seen in coastal areas. In the interior, it usually occupies south-facing mountainous slopes and is most common on shales but also found on sandstones. The plants occur in mountain fynbos conditions with a relatively high rainfall. Most flowering appears to occur in the first few years after a fire, after which the plants flower rather poorly. Plants of the smaller forms coppice from the rootstock after a fire, but those of the tall forms are killed by fire and regenerate from seed. The rhizomes form dense stands and the plants can reach a diameter at ground level of about 2 m and a crown diameter of up to 3 m. The strong stems or culms have clusters of sterile branches at the nodes that can reach a length of 1 m and give the plants a very graceful appearance. The inflorescences grow up to 0.5 m long and the male inflorescences (left, background and photo below) are pale gold in colour and resemble a long raceme with short side branches with small, golden flowers. The female inflorescences (left, foreground) are much shorter and have tight pale brown bracts that conceal small insignificant flowers. After pollination, it takes nearly a year for the seeds to ripen.

Cannomois virgata is one of the nut-fruited restios with indehiscent fruits that have a hard woody ovary wall. Seed germination is promoted by smoke treatment and by incubating seeds at alternating temperatures of 18°C and 28°C. Seedlings grow fairly fast and attain their final height 2–3 years after planting out. It is ideal for mass plantings or as an accent plant in large gardens. The young shoots are ideal for flower arrangements and the female inflorescences are most ornamental and highly prized in the dried flower trade. Sow seeds from late summer to early autumn.

Chondropetalum tectorum

Common name: dakriet

Height: small form 1-1.5 m

Flowering time: late summer to early autumn

A handsome, vigorous species found from Clanwilliam in the Western Cape to Port Elizabeth in the Eastern Cape, growing in large stands in damp localities. It has a tufted growth form and reaches a spread of 1.5-3 m. The plants have slender compact flowering spikes and it has a fast growth rate with a juvenile growth stage of 1.5 years and a lifespan of approximately 10 years. The specific name *tectorum* means 'roofing' and suggests that it was once used as thatching material. An excellent species for poorly drained areas in full sun and windy seaside gardens, and is tolerant of alkaline soil and light frost. The female plant is shown above. Sow seeds in late autumn.

Elegia capensis

Common names: besemriet, fonteinriet

Height: medium form, 1-1.5 m; tall form, 2-3 m

Flowering time: spring to early summer

A very attractive species that occurs in at least two growth forms, one reaching from 1-1.5 m in height and another growing to a height of 2-3 m. The plants grow in clumps or tussocks and have a spread of up to 1.5 m. They have slender branches arranged in whorls at the nodes, similar in appearance to *Equisetum* (horsetails). Plants produce golden brown flowers for about 3 weeks and the seeds ripen in late summer. In the wild this species is common along seepage areas in mountain fynbos in the Western and Eastern Cape. It is long-lived and a fast grower, and plants can reach a height of 1 m in the first year from seed. Seeds germinate relatively easily. Highly recommended for full sun positions in large gardens with poorly drained, acid soils. The male plant is shown above. Sow seeds in late autumn.

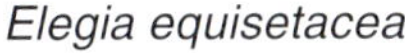

Elegia equisetacea

Height: 1.5 m

Flowering time: autumn to early summer

A highly ornamental, reed-like plant with finely divided side stems in tufts along the main stems. Seedlings reach a height of about 1 m during the second year after sowing and will have reached their full height after flowering in the third year. In their natural environment the plants are relatively long lived, and in gardens with richer soils and a more generous supply of water, they will live at least 7 years. They require full sun, a well-drained acid soil and plenty of air movement. Suitable for use in a mixed fynbos border, for landscaping purposes or as specimen plants for growing in large containers. Plants are best grown from seeds which germinate well when pre-treated with smoke or a commercial smoke seed primer. Seeds take about 3 weeks to germinate. The male inflorescences are shown above in front of the females. Sow seeds in late autumn.

Elegia persistens

Height: 0.3-1 m

Flowering time: late summer to early autumn

A most attractive, compact, medium-sized tufted perennial with spreading rhizomes, occurring naturally in mountainous terrain in the southwestern and southern Cape. The flowers of the female inflorescences (see foreground above) are completely enclosed by persistent golden spathes that remain attractive for several months, hence the specific name persistens. The male inflorescences are pale to dark brown and feathery (seen behind female plant above), creating excellent textural and colour contrast. The plants require acid soil in full sun, moist conditions in winter and are waterwise once fully established. It is an excellent subject for small or medium–sized gardens in mixed fynbos beds or rock garden pockets, and can also be grown successfully in large containers. Sow seeds in late autumn.

Ischyrolepis subverticillata

Common name: besemriet

Height: 2 m

Flowering time: late summer to early autumn

This species occurs naturally from Paarl to Caledon, where it grows in sunny positions in seasonally wet riverbeds. Plants may be found as close as 50 m from the sea, to positions in light shade along small streams high up in the Kogelberg Nature Reserve. It is thus a plant for full sun or semi-shade, and sandy, well-drained, acid soils, preferably in moist situations. Mature plants eventually form large clumps 2 m high and more than 3 m in diameter, and can be used as accent plants or in groups. Female flowers develop into beautiful shiny speckled grey nutlets and the male flowers (above) are borne on attractive, feathery inflorescences. Its main decorative value lies in its sprays of dark green feathery foliage that are used in the cut-flower industry. Highly recommended for large gardens. Sow seeds in late autumn.

Restio dispar

Height: 1-2 m

Flowering time: autumn

A tufted, tall growing species from the southwestern Cape, occurring from the Cape Peninsula to Caledon, and inland to Worcester. The female plants (above) are much more showy, sturdier and larger than the male plants and have long, reddish-brown inflorescences while the males have smaller, paler inflorescences. Plants can reach a diameter of about 1 m at ground level after about 3 years of growth. Seed germinates readily and even better if pre-treated with smoke or a commercial smoke seed primer. It is a fast growing plant that will reach about 1 m in height a year after sowing, and will have formed a handsome plant within 3 years. This waterwise plant is an ideal subject for large gardens and is especially well suited to planting on sloping ground. Sow seeds in late autumn.

Jeanette Loedolff

Rhodocoma gigantea

Common name: dekriet

Height: 2 –3 m

Flowering time: winter

A spectacular tall restio found on seasonally wet mountain slopes. During their first two years the seedlings produce a mass of finely branched, bright green juvenile foliage. During the third year they reach their full height with striking golden brown inflorescences. The female flowers (see foreground above) are erect and clustered at the nodes while those of the male plants (see background above) are heavily drooping and feathery. The plants require full sun, deep acid soil and moist conditions in winter, and are wind tolerant. They are suited to medium-sized or large gardens, and reach an age of at least ten years. Seeds should be sown in late autumn and germination is improved by pre-treatment with smoke or a commercial smoke seed primer (see page 44).

Thamnochortus insignis

Common names: dekriet, thatching reed, Albertinia thatching reed

Height: 2.5 m

Flowering time: summer

This outstanding restio comes from the southern Cape. It forms dense tufts with a spread 3-4 m wide, and has heavily lignified stems that are widely used for thatching. The female inflorescences have numerous erect, cone-shaped spikelets while those of the male inflorescences (above) are hanging and feathery. The plants have a fast growth rate, with a juvenile period of 1.5 years and a lifespan of about 30 years. It thrives in alkaline, sandy soils and tolerates sea wind and light frost but performs equally well in a range of different soil types, and is very waterwise. Germination takes approximately 4 weeks and pre-treating seed with smoke or a commercial smoke seed primer improves germination. Sow seeds in late autumn.

Opposite: *Chondropetalum tectorum* (female plant) is an excellent species for poorly drained soils in full sun (see page 50)

Below: *Elegia stipularis* (male plant) at the Cape of Good Hope Nature Reserve (see page 188), with the golden conebush *Leucadendron laureolum* in the background (see page 31)

Thamnochortus spicigerus

Common names: dekriet, olifantsriet

Height: up to 1.5–2 m

Flowering time: autumn and winter

A handsome, robust species occurring in large colonies in deep coastal sand from Aurora to the Cape Peninsula. The plants form thick clumps that increase in size by means of strong rhizomes. Like the similar-looking *Thamnochortus insignis*, the heavily lignified stems are used for thatching. The male plants (above) produce inflorescences consisting of numerous spreading, feathery, hanging spikelets while those of the female plants are erect and cone-shaped. It is a long-lived species highly recommended for hot and dry, windy coastal gardens, in alkaline or acid soils. The ripe seeds should be harvested in winter and a marked improvement in germination is obtained when treated with smoke or a commercial smoke seed primer (see page 47). Sow seeds in late autumn.

Background photograph: Display bed in the Restio Garden at Kirstenbosch

 These two pages kindly sponsored by Silverhill Seeds

Elegia capensis
Thamnochortus insignis
ata
Thamnochortus spicigerus
Ischyrolepis subverticillata

Below: The wax heath *Erica ventricosa* was grown in England in the 19th century, long before it became known in cultivation in South Africa (see page 189)

Right: *Erica versicolor* (see page 71)

Alice Notten

4. GROWING FYNBOS ERICAS

(Compiled from text originally written by Anthony Hitchcock, Deon Kotze, Alice Notten, Inge Oliver and Ted Oliver)

Erica is the largest genus in the South African flora with 760 species. There are about 860 species in the world, occurring from the Cape Peninsula up through Africa and also in Madagascar, to Europe where 21 species are recorded.

Ericas are one of the 3 main constituents of the famous and very distinctive Cape Flora or fynbos that dominates the mountains and the coastal flats of the Western Cape. Ericas belong to the worldwide family Ericaceae that contains such well-known plants as rhododendrons and azaleas, bilberries, blueberries, cranberries and the Australian heaths (epacrids).

What makes an erica distinct from other members of our flora? Most of the *Erica* species have needle-like 'ericoid' leaves, each with a narrow channel beneath. The leaves are borne in whorls around the stem, usually in threes or fours. The petals are joined together to form a corolla tube that may be narrow or open and cup-shaped. Ericas are perennial shrubs and on average are 300–500 mm tall. Some can be large shrubs to small trees up to

Opposite: *Erica formosa* (see page 189)

Above: The Elim heath *Erica regia* is highly recommended for rock garden pockets and containers, and is a useful cut-flower (see page 70)

4 m tall, while others can be prostrate, spreading among other vegetation, or very small plants a mere 10 mm tall and only 30 mm across.

Most ericas grow on open, sandy stony slopes and flats. However some are adapted to grow as moss-like plants only on moist, often shady, rock ledges. The size of the flowers varies considerably from the very showy long-tubed species up to 45 mm long, to minute species with flowers only 0.7 mm long. Much of the size, shape and colour of the flowers is linked to its method of pollination. The long-tubed species are pollinated by sunbirds, the general urn-, bell- and cup-shaped species by various insects and the minutely-flowered, dull-coloured species by wind.

The family Ericaceae in southern Africa used to be divided into 23 genera. There was the very large genus *Erica,* and the rest were minor genera, but the diagnostic features used to distinguish these are no longer regarded as valid and all species are now included in one genus - *Erica.*

Cultivating ericas

There are over 700 species of *Erica* of which at least 50 have good garden potential. Their appeal lies in the varied form of their flowers. The blooms come in a multitude of colours, shapes, sizes and arrangements such as shiny long tubes, inflated star-faced tubes or an extravagant profusion of little bells. The tube-shaped flowers attract nectar-loving birds. In the winter rainfall region, gardening with ericas is not difficult. The natural climate and soils allow for a great choice of species, although it is nevertheless still important to select appropriately tough species that

will survive in your garden. Ericas require a sunny position, well-drained soil, low phosphate levels and regular watering.

In the summer rainfall regions, there is no sense in trying to grow all the attractive species available to Cape gardeners. The golden rule is to make sure that the species you buy are known to be suitable for Highveld conditions. Plant ericas in a sunny position where they get good air circulation. If the garden is south-facing and the situation allows a sunny position, a protecting wall or hedge is advisable to ward off cold wind. In such conditions growing ericas in containers would perhaps be more suitable, as pots can be moved to more sheltered positions in winter.

It is important to plant ericas in the correct soil. The potting medium should be well-drained and acidic, contain no manure, and have low levels of phosphate. Well-drained, sandy loam with a pH of between 5 and 5.5, containing approximately 50% humus, is recommended.

Erica viridiflora is a long-lived species under cultivation, best suited to rock garden pockets (see page 189)

Successful planting

Ericas grow particularly well in rockeries or sloping ground, but level sites are also suitable. Dig over the ground to loosen and aerate the soil and to clear it of grass and weeds. Dig in pure compost (preferably pine needles) or decomposed pine bark. Make sure that the compost is well rotted before planting, otherwise the roots may be burnt. Avoid manure-based or mushroom composts.

The ideal time to plant in the Cape and milder areas on the Highveld is in autumn (March, April and May). The cooler weather will assist rapid development of the root system, as well as top growth. In areas with colder winters, ericas may be planted in August, September and October. It is not advisable to plant in summer. Ericas have delicate root systems that are susceptible to disturbance and drying out. The plastic bag or pot should be removed carefully and the root-ball kept intact while planting in a pre-prepared hole. The planting hole should be just large enough to accommodate the size of the container. Fill in around the plant, firm the soil down lightly and then water well. Place a thick mulch of decomposing pine needles or untreated wood chips around the plant. Mulching keeps the soil surface around the plant cool and suppresses weed growth.

Ericas and fynbos garden design

Ericas grow better when planted close together with other fynbos plants to form dense stands that cover the ground. This approach will help to suppress weed growth and keep the soil cooler. Spacing will depend on the size and growth form of the selected plants. Mixing ericas with other plants is a good practice because it helps to limit the spread of fungal diseases that are specific to one plant group.

Ericas can dominate a garden when they are in flower, so select an erica for every season to provide some colour all year round. Consider also the colours and textures of the foliage of the plants you choose and this will make the garden an interesting and attractive place throughout the year.

Choose a dominant or backbone plant group to hold the textural effect together. The Cape reeds or restios are ideal for this purpose. You should also look for silvery foliage colours of helichrysums or similar plants and use them as fillers. If you want to create a natural-looking fynbos garden, combine ericas with proteas, buchus, phylicas, wild daisies and bulbs.

Caring for ericas

Ericas can be lost through lack of proper care, especially from irregular or inadequate watering. A good soaking is essential to ensure that the water penetrates well into the soil. Look out for signs of water stress (shown by drooping stem tips) even in the rainy season. In the winter rainfall region, ericas should be watered during the summer to keep them in good condition. In the summer rainfall area, ericas should be well watered during the dry winter months. In addition watch out for dry and very hot spells in summer and give additional water if necessary. If the water in your area is alkaline or 'brak' it may be advisable not to grow ericas at all.

Water in the morning when it is still cool. Watering in the heat of the day may result in scorching. If the plants are wet at night, fungal infections may develop on the leaves. A good rule is to water well every second or third day. Chlorine in water may gradually disturb the ideal soil condition particularly in areas with a low rainfall. The application of an acidifying agent, such as ammonium sulphate will help to keep the soil pH low.

Ericas respond very well to the correct fertilizers. Organic liquid foods derived from seaweeds and Seagro fish emulsion have traditionally been the safe method of fertilizing this genus. You can also use controlled-release granular fertilizers, such as Osmocote but make sure that the fertilizer is low in phosphates. These fertilizers release small amounts of nutrients continuously and in the correct quantities. The ones recommended for azaleas can also be given to ericas. Mix the granules into the soil according to the advised dosage.

Regular pruning will improve the shape of the plant and result in greater flower production. This is also an opportunity to cut away any diseased parts of the plant. The best time to prune is shortly after the plants have finished flowering.

Ericas are not immune to diseases and these can create havoc in a fynbos garden. However, the disease usually only takes hold because of poor growing practices. If careful attention is paid to the basic requirements for growing ericas, then the stresses on the plants can be kept to a minimum and the plants should remain healthy. Control through the use of fungicides is difficult and expensive and, by the time the plant shows symptoms, it is often too late.

A quick guide to cultivating ericas

- Plant in sunny positions
- Plant in well drained, acidic soils containing well-rotted compost
- Avoid warm, humid conditions, such as tropical areas
- Allow free air movement around plants
- Keep soil surface around roots cool by mulching
- Avoid disturbing their roots by digging
- Protect from high temperatures over long periods
- Water regularly
- Protect from severe cold or frost
- Use fertilizers low in phosphates
- Apply small amounts of nutrients regularly
- Prune plants regularly to improve shape and flowering
- Prune and remove diseased material

Vegetative propagation

Growing ericas from cuttings is faster than growing them from seed.

Vegetative propagation is done with mist and bottom heat and good air circulation is important, as it makes the cuttings less susceptible to fungal infection. As a general rule, the time to take cuttings is approximately 2 months after flowering. At this stage the plants will have grown 50–100 mm. Plants must be healthy and free from disease. The cuttings are taken from semi-hard wood, and should be 40–50 mm long. Heel cuttings are most suitable although nodal cuttings may also be used.

The leaves on the lower third of the cutting are removed. Rooting hormones can be applied with beneficial results. Rooting media vary considerably but the best results have been achieved with 50% peat or crushed pine bark and 50% polystyrene. Bottom heat is used and kept

Erica baccans (see page 67)

at a temperature of 22°C–24°C. Once the cuttings are well rooted, they are potted into half-litre plastic bags using a fynbos growing medium.

The young cuttings are watered well after potting and shaded lightly for a month. The bags are then placed in full sun. Care must be taken not to damage the fine root hairs when transplanting. Within 3 to 4 months the young cuttings are ready to be planted out.

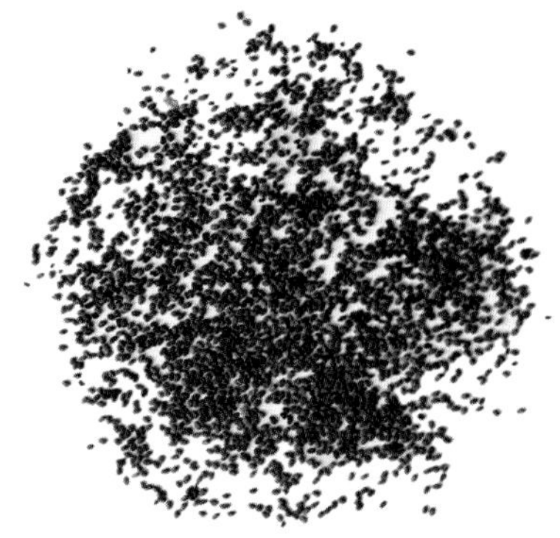

Seeds of *Erica baccans*

Propagation from seed

Seed should be harvested just as it starts to fall naturally. The old flowers containing the fruits should be dried and then rubbed through a sieve. The resultant mix may be winnowed to separate the seed from the chaff. The seed is very fine and often sticky.

The seed should be sown in late autumn (April-May) in a seed tray not less than 100 mm deep. The sowing medium should be well drained and acidic, and be firmed down to give a level surface. If this surface is not level, germination is not even. Prior to sowing, the seed tray should be well watered with a fine rose. Seeds may be pre-soaked in aqueous smoke extract or a commercial smoke seed primer for 24 hours before sowing; or seeds may be smoked once sown in trays. An old tea or coffee strainer is very useful for use in handling very small seeds. Seeds are placed on the gauze of the concave side of the strainer and this is dipped into the smoke solution. After 24 hours the strainer is taken out of the solution and the excess liquid is drained off and the seeds are dried.

Once pre-soaked seeds have been dried off, they are then sown evenly over the whole surface of the seed tray. Care must be taken to obtain an even distribution, thus minimizing the chances of damping-off. The seed can be mixed with fine sand to obtain better distribution. A fine layer of the sieved growing medium can be placed over the sown seed, although this is not necessary. Water gently with a fine rose, and keep the seed tray out of direct sunlight and rain in an area with good air circulation. Germination time varies from 1–2 months.

When the seedlings are approximately 10 mm tall, place them in the open, under light shading. During the period from October to December, they will have reached a height of 20–50 mm and will be ready for pricking out and planting in a fynbos potting medium. As many of the fine hair-like roots as possible must remain on the plants. The seedlings should then be placed under light shade and watered well. Once established, shading is not required, with the exception of areas where the temperature exceeds 32°C.

In the wild, fire is very important in the ecology of ericas and the vast majority of species regenerate only from seed after a veld fire. Seeds are very small and, in all but one species, are shed when ripe. Serotiny is rare in this family and is found only in *Erica sessiliflora*.

A quick guide to germinating *Erica* seeds

- Use fresh, mature opaque seed
- Soak seeds in aqueous smoke extract or commercial smoke seed primer for 24 hours before sowing; or smoke seeds sown in seed trays. The trays should have a sand/loam/bark mixture and be well drained
- Alternatively, seeds may be pre-soaked in gibberellins GA_3 or GA_4 and GA_7 solution prior to sowing
- Optimum seed germination is obtained if seeds are incubated under alternating night/day temperatures, e.g. 10°C (16 hours night); 15°C–25°C (8 hours day) as would be found in open areas of burnt fynbos in the autumn or winter

The following *Erica* species are suitable for both Highveld and winter rainfall conditions (See also, 'Ten of the Best' starting on page 67).

Spring-flowering species include *Erica abietina* subsp. *constantiana, E. baueri* subsp. *baueri* (Albertinia heath), *E. blenna* (Chinese lantern heath), *E. chamissonis* (Grahamstown heath), *E. cyathiformis, E. fastigiata* (four sisters heath), *E. holosericea, E. mauritanica, E. nana, E. patersonii* (mealie heath), *E. peziza* (kapokkie), *E. viscaria* subsp. *gallorum* and *E. walkeria* (Swellendam heath).

Spring- and summer-flowering species include *Erica caffra* (water heath), *E. formosa, E. glauca* var. *glauca, E. gracilis, E. lanipes, E. mauritanica, E. nutans, E. quadrangularis* (baby heath), *E. scabriuscula (=E. gibbosa), E. sparrmannii* and *E. tenella*.

Autumn-flowering species include *E. baccans, E. patersonii* (mealie heath), *E. sitiens* and *E. verticillata*.

Winter- and spring-flowering species include *E. baueri* subsp. *baueri* (Albertinia heath) and *E. sparsa* (ker-ker).

Summer-flowering species include *E. densifolia, E. hirtiflora, E. mammosa* (ninepin heath), *E. sitiens* and *E. ventricosa* (wax heath).

Species that flower all year, on and off, include *E. cerinthoides* (fire heath), *E. coccinea, E. curviflora, E. glomiflora, E. hebecalyx* and *E. versicolor*.

Erica blenna (see page 68)

Ten of the Best Fynbos Ericas

(see also Appendix 2, page 189)

Erica baccans

Common names: berry heath, bessieheide

Height: 2–3 m

Flowering period: autumn to early spring

This long-lived, erect bushy shrub is endemic to the Cape Peninsula and becomes covered in masses of small rose-pink, berry-like flowers. It has a very long flowering period and is an excellent companion plant in a mixed fynbos garden. It thrives in moist but well drained acid soil and should be pruned back lightly after flowering to maintain a compact shape and prevent plants from becoming too lanky. An excellent cut-flower. Sow seeds in late autumn.

Anthony Hitchcock

Erica baueri subsp. *baueri*

Common names: Albertinia heath, bridal heath, witheide

Height: 1–1.5 m

Flowering time: winter to spring

A large, erect shrub with distinctive grey-green leaves that produces masses of white or pink tubular flowers near the ends of long thin stems. It requires acid soil and is best planted in groups for maximum effect. A good cut-flower, and one of the most popular and widely cultivated of the South African ericas. It is fairly hardy and long-lived in cultivation. Easily grown from seed, but propagation from cuttings is more difficult. Fairly waterwise and highly attractive to sunbirds. Sow seeds in late autumn.

Erica blenna

Common names: Riversdale heath, Chinese lantern heath, belletjieheide

Height: up to 1 m

Flowering time: autumn to early summer

One of the most striking of all the ericas, it forms an erect small shrub with slender branches and bunches of eye-catching orange, urn-shaped sticky flowers tipped with dark green. It has a limited distribution in the southern part of the Western Cape and there are small- and large-flowered forms, both of which are recommended for cultivation. The branches should be pruned back lightly after flowering to prevent plants from becoming too woody. It is well suited to cultivation in rock garden pockets in acid soil and also makes a charming container subject. Sow seeds in late autumn.

Erica cerinthoides

Common names: fire heath, rooihaartjie

Height: 0.2 –1.2 m

Flowering time: almost all year round

A very showy small shrublet with willowy branches. It has a persistent rootstock that enables the plant to regenerate after a fire. It has tubular hairy, sticky crimson-red flowers in clusters at branch tips. It is a tough plant that grows easily in different parts of South Africa in acid soil and can live for a long time in the garden. Suitable for beds, rock gardens and containers and can be kept compact and well shaped by light pruning after flowering. Not hardy to severe frost but should re-sprout if damaged by frost. Easily grown from seed sown in late autumn, or from cuttings.

Erica coccinea

Common name: small tassel heath

Height: 0.8–1.2 m

Flowering time: throughout the year, but mainly in late winter

An erect, robust, stoutly branched shrub. Its striking tubular, pendulous, yellow to red flowers are borne in closely packed bunches near the tips of branches and have conspicuous long brown anthers protruding well beyond the perianth tips. It is a long-lived plant in cultivation and one of the easiest ericas to grow, highly recommended for rock garden pockets in acid soil. Sow seeds in late autumn.

Brenda Szabo

Erica lateralis

Common names: button heath, knopiesheide

Height: 0.4–0.6 m

Flowering time: midsummer to late autumn

A low-growing bushy shrub from the southwestern Cape producing a spectacular show of bright pink, urn-shaped flowers in bunches at the branch tips. It is best planted in groups for maximum effect and is well suited to rock garden pockets and containers in acid soil. Light pruning of the branch tips after flowering keeps the plants compact and well shaped. Sow seeds in late autumn.

Erica patersonii

Common names: mealie heath, mielieheide

Height: 0.6–1 m

Flowering time: autumn to early summer

A slender, erect, sparsely-branched small shrub with sub-erect, needle-like leaves and dense spikes of tubular bright yellow flowers resembling mealie cobs. It is a floriferous and easily cultivated species, highly recommended for rock garden pockets in acid soil, best planted in groups for maximum effect. Old spikes should be pruned back to encourage branching and prevent plants becoming too lanky. It likes moist conditions and is a good cut-flower. Sow seeds in late autumn.

Erica regia

Common name: Elim heath

Height: up to 0.7 m

Flowering time: throughout the year, with a peak in early spring

A well-known erica and a popular garden subject due to its striking flower colouring and relative ease of propagation. The corolla is tubular and sticky, and flowers occur in a range of colours. Particularly favoured for cultivation is the colour variant with white corollas and red tips, sometimes merging from white into green, then red. The variant that is uncommon in the wild and in cultivation is a uniform crimson red. If left to their own devices the plants may reach 0.6–0.7 m in height, but in so doing will become rather woody and lanky, but if the branches are pruned back lightly after flowering a more compact shape can be maintained. Recommended for rock garden pockets and containers in acid soil, and a useful cut-flower. It is also highly attractive to sunbirds. Sow seeds in late autumn.

Erica versicolor

Height: 1–3 m

Flowering time: all year round

An attractive, robust and long-lived, erect or sprawling shrub. The large, sticky tubular flowers are borne in clusters and are variable in colour, ranging from pink with paler pink tips to dark red with green or yellow tips. One of the most easily grown ericas, it is waterwise and suited to both small and large gardens in acid soil. It is a good cut-flower and highly attractive to sunbirds, and an essential component of every fynbos garden. Sow seeds in late autumn.

Erica verticillata

Height: 1–2 m

Flowering time: midsummer to late autumn

This beautiful erica is extinct in the wild. It is a strong grower that responds very well to cultivation. It requires acid soil and is long-lived and very floriferous, forming a fairly dense shrub bearing clusters of large tubular, mauvish-pink flowers. Light pruning of the branches after flowering is recommended to prevent plants from becoming too lanky. It is best planted in groups for maximum effect in the garden, and is also suited to rock garden pockets and as a container plant. It is fairly waterwise and highly attractive to sunbirds. Sow seeds in late autumn.

Below: *Chrysanthemoides monilifera* is waterwise and an excellent subject for stabilizing sandy soil in windy seaside gardens, and readily adapts to any soil type (see page 75)

Opposite: *Dimorphotheca cuneata* (see page 76)

5. GROWING FYNBOS DAISIES

The Asteraceae, usually the largest family in floras of arid to semi-arid regions, is also the family with the most species (1036, including 655 endemics) in the Cape Flora, many of them of horticultural importance. Amongst these are the 'everlastings' with their colourful inflorescences that are harvested and dried for the dried wildflower industry.

Propagation of fynbos daisies

Seeds are produced from the fertile flowers at the centre of the flower heads, the sterile flowers on the periphery provide the colourful papery bracts. In nature the seeds of the 'everlastings' in the Asteraceae germinate in response to the smoke of a veld fire, and large numbers of seedlings then appear. The plants then gradually die out, until a flush of germination occurs after the next fire.

A quick guide to germinating seed of the Asteraceae

- Use only mature, plump, fully formed seeds. In most species these are darker in colour than the immature ones
- Pre-soak seeds in aqueous smoke extract or a commercial smoke seed primer for 24 hours or smoke seed trays after sowing

- Seeds should be give a light dusting with a fungicide dressing to prevent post-emergence seedling infection
- Sow seeds in a sandy, well-drained soil medium. Incubate in full sun under autumn temperature conditions, e.g., alternating 10°C (16 hours night) x 20°C (8 hours day)
- Seeds of some species are sensitive to light. Light may either promote or inhibit germination. There is also an interaction between the effect of light and the germination response to smoke

Below: *Dimorphotheca cuneata* is frost-hardy and waterwise (see page 76)

Ten of the Best Fynbos Daisies

(see also Appendix 2, page 190)

Arctotis revoluta

Common name: krulblaargousblom

Height: 1–2 m

Flowering time: early spring to early summer

The curled, greenish-grey foliage of this small rambling shrub emits a distinctive, unusual aroma when bruised. It produces large, solitary bright yellow or orange flower heads on long slender peduncles and is highly recommended for rock gardens and dry, difficult coastal gardens. It performs best in well drained loamy soil but easily adapts to a range of soil types. The branches should be pruned back lightly after flowering to maintain a compact, neat appearance. Sow seeds in autumn.

Arctotis stoechadifolia

Common names: silver arctotis, kusgousblom

Height: up to 0.3 m

Flowering time: spring to summer

A trailing perennial with finely divided, silvery-grey leaves and large cream-coloured flowers shading to brownish-purple on the reverse. An excellent waterwise groundcover for stabilizing steep banks and adapts to any soil type. Easily propagated by cuttings and division of rooted, trailing stems in autumn and spring, and especially well suited to difficult coastal gardens. Sow seeds in autumn.

Chrysanthemoides monilifera

Common names: bietou, bosluisbessie

Height: 0.7–1.8 m

Flowering time: late summer to midwinter

A spreading, dense rounded shrub with dark green leathery leaves and masses of bright yellow flower heads borne in terminal corymbs on short peduncles. The abundance of bright green berries produced directly after flowering mature to black and are relished by numerous fruit-eating birds. It is an excellent subject for stabilizing sandy soil in difficult, windy seaside gardens and readily adapts to any soil type. It is also very waterwise. Sow seeds in autumn.

Dimorphotheca cuneata

Common name: gousblom

Height: 0.3–0.6 m

Flowering time: early spring to early summer

A rounded bushy perennial with hairy, divided leaves and large white and yellow flower heads borne singly on short to long slender peduncles. The darker coppery undersides of the ray florets are especially attractive and it is a fast-growing plant ideally suited to well drained rock garden pockets in full sun. The plants should be pruned back after flowering to encourage new, strong growth. They require well-drained loamy soil, regular watering in winter, have good frost tolerance down to -3°C, and are waterwise. Sow seeds in autumn.

Eriocephalus africanus

Common names: wild rosemary, wilderoosmaryn, kapokbos

Height: up to 1 m

Flowering time: varies, with best displays in winter

An excellent spreading, evergreen shrub with attractive fragrant, linear, silvery-grey leaves. The small white flowers are borne in clusters at the tips of branches and the fruits mature to attractive fluffy seed heads (above) that resemble cotton wool or snow, hence the descriptive Afrikaans common name 'kapokbos'. The plants prefer full sun and well-drained soils, and are very waterwise. They should be pruned back lightly after flowering to encourage bushy growth. Their extensive root system makes them highly resistant to drought and able to recover from animal grazing. Easily propagated from seed sown in autumn or from tip or heel cuttings taken in autumn or spring.

Liesl van der Walt

Euryops virgineus

Common names: honey euryops, heuningmagriet

Height: up to 3.5 m

Flowering time: late winter to spring

A striking, densely branched evergreen shrub with a spread of up to 1.5 m. Masses of small, bright yellow, honey-scented flower heads 8–10 mm in diameter are borne singly on short peduncles in the upper portion of the branches, and are visited by large numbers of honey-bees. These fast-growing plants require full sun and are frost-hardy and drought- and wind-resistant when established. They are especially useful in new gardens, rapidly filling empty spaces and are excellent for screening purposes. They need to be pruned back hard after flowering to prevent them from becoming too lanky. Easily propagated from seed sown in autumn or cuttings taken in autumn or spring.

Felicia aethiopica

Common name: wilde-astertjie

Height: 0.5–1 m

Flowering time: all year round but mainly in spring

A spreading, small compact shrublet with bright green elliptic foliage and eye-catching, solitary bright blue (or white) and yellow flower heads borne on long slender peduncles. It is ideally suited to containers, rock garden pockets, as an edging plant to taller perennials, or allowed to sprawl over low garden walls. Easily propagated from seed sown in autumn, or cuttings taken in autumn or spring, it requires well drained soil in full sun and easily adapts to new soil types. Recommended for difficult coastal gardens and is waterwise.

Felicia echinata

Common name: bloublommetjie

Height: 0.6–1 m

Flowering time: autumn to early summer but mainly in late winter

An attractive small shrub with erect branches densely covered with dark green, leathery lance-shaped leaves and numerous large mauve (or white) and yellow flower heads. It is a very floriferous, easily grown plant, ideally placed towards the centre of herbaceous borders or in rock garden pockets. It thrives in full sun in moist loamy soil and the branches should be pruned back heavily after flowering to encourage strong new growth. Easily propagated from cuttings taken in autumn or spring, and waterwise. Sow seeds in autumn.

Gazania krebsiana

Common names: oranjegousblom, rooigazania

Height: up to 0.2 m

Flowering time: spring to summer

A very variable, clump-forming perennial with magnificent bright orange, red or yellow flowers. It is ideally suited to rock garden pockets or for massed planting towards the front of herbaceous borders. Plants require full sun and a well-drained soil to flower well and are resistant to summer drought. Well suited to inter-planting with fynbos bulbs and low-growing mesembs, and an excellent choice as a waterwise plant. The equally attractive and variable *Gazania pectinata* (below) is also highly recommended. Sow seeds in autumn.

Heterolepis aliena

Common names: rock daisy, rotsgousblom

Height: 0.3–0.6 m

Flowering time: spring to midsummer

A spreading, densely leafy shrublet with a sprawling habit and a woody rootstock. The leaves are long and narrow and the spectacular bright yellow flower heads are about 60 mm in diameter and held on short, rough hairy stalks. It requires sunny conditions in well-drained soil and is suited to fynbos gardens, waterwise gardens, rock gardens, slopes, banks, terraces and retaining walls. The plants like slightly acidic soil, are sensitive to over-watering and benefit from a compost mulch applied in autumn and spring. They are easily propagated from seed sown in late summer or autumn, and heel cuttings taken in autumn or spring. Softwood cuttings should be rooted under mist with bottom heat, while harder growth can be rooted in a cold frame.

Below: *Euryops virgineus* is especially useful in new gardens, rapidly filling empty spaces (see page 77)

Below: *Arctotis hirsuta* (orange) and *Dimorphotheca pluvialis* (white) form a carpet of colour along the Cape West Coast near Langebaan (see page 83)

Opposite: A yellow form of *Dorotheanthus bellidiformis* (see page 84)

6. GROWING FYNBOS ANNUALS

An annual is a plant that completes its life cycle in one year. In other words, seeds germinate and grow into adult plants that flower, produce seed and then die, all in one growing season that consists of about 7 months. In this way the unfavourable time of the year, the hot dry summer or the cold dry winter, is survived safely in the form of seed.

Spring annuals

Every year in spring the Western Cape has an annual display like nowhere else in the world. All along the west coast from Alexander Bay to Cape Town there are wildflower displays of pure magic. Fields of annuals cover the ground in a spectacular range of colour from brilliant white, soft peach, intense orange and yellow to striking blue and mauve.

It is mostly the daisies (members of the Asteraceae) like *Dimorphotheca*, *Arctotis, Felicia, Osteospermum* and *Gazania* that create the sheets of colour. When looking more closely, many other annuals like *Diascia, Nemesia* (both belonging to the Scrophulariaceae), *Heliophila* (Brassicaceae), *Wahlenbergia* (Campanulaceae) and *Grielum* (Neuradaceae) can be identified.

Many of these annuals grow well in ordinary gardens. The time for sowing spring-flowering annuals is in autumn (March-April). The seeds can be sown by broadcasting directly in a prepared flowerbed and raked into the soil in summer rainfall areas. In winter rainfall regions weeds are a problem in the garden so it is preferable to sow in seed trays or open seed beds, from which they are planted out into the garden. Most of the annuals need a light soil with good drainage and full sun.

Seed sowing

Seed can be pressed into sandy soil or covered with a light sprinkling of soil - just enough to cover the seed. Fluffy, featherweight seeds, such as *Arctotis* and *Ursinia* (both Asteraceae) should be pressed firmly into the surface of the soil and covered to prevent them blowing away. If sowing is done in very dry weather, the ground should be well watered beforehand and kept reasonably damp.

Do not sow seeds thickly. In the case of the species with tiny seeds, mixing the seeds with a little dry sand promotes even sowing.

Garden use

For an effective display, annuals should be planted *en masse* or mixed with fast growing herbaceous plants. After flowering, the plants can be left until the seed has ripened before they are removed to make way for summer annuals. The seed collected in the garden may then be used for the following year's spring annuals.

Southern African annuals are primarily lovers of the sun and therefore need to be grown in open situations with full day sun. The soil should be well drained, light and friable.

Below: *Senecio elegans* (mauve) in the Tinie Versfeld Wildflower Reserve near Darling (see page 87)

Ten of the Best Spring Flowering Fynbos Annuals

(see also Appendix 2, page 190)

Arctotis hirsuta

Family: Asteraceae

Common name: gousblom

Height: 0.3–0.5 m

Flowering time: spring

A slightly fleshy, often robust annual with soft hairy leaves. The flower heads are blackish with yellow, orange or cream ray florets that have a dark band at the base. Easily raised from seed sown in autumn, they can be grown in both sandy or heavy soils, as long as they are well drained. Suitable for inter-planting with larger fynbos bulbs like *Chasmanthe floribunda* and *Watsonia borbonica*.

Dimorphotheca pluvialis

Family: Asteraceae

Common names: white Namaqualand daisy, reenblommetjie

Height: 0.2–0.3 m

Flowering time: late winter to spring

A very floriferous erect to sprawling annual with toothed and hairy, lance-shaped leaves. The solitary flower heads are purple with brilliant white ray florets that have a narrow to broad purple band at the base and attractive darker undersides. Prefers sandy soil but easily adapts to heavier soil types provided they are well drained. Easy to grow from the papery, disc-like seeds that should be sown in autumn.

Liesl van der Walt

Dimorphotheca sinuata

Family: Asteraceae

Common name: Namaqualand daisy

Height: 0.3–0.4 m

Flowering time: spring

An erect to sprawling, slightly hairy-leafed annual with solitary flower heads borne on elongated wiry peduncles. The flower heads are dark yellow with paler yellow, orange or biscuit-coloured ray florets. Grows equally well in sandy or heavy, well drained soils in full sun, and suitable for inter-planting with medium-sized fynbos bulbs like *Moraea elegans* and *Watsonia laccata*. Easy to grow from the papery, disc-like seeds that should be sown in autumn.

Dorotheanthus bellidiformis

Family: Aizoaceae

Common names: Bokbaaivygie, Livingstone daisy

Height: 0.04–0.06 m

Flowering time: spring

A succulent, low-growing tufted annual with small glistening leaves and masses of brilliant flowers on slender stalks in purple, yellow, salmon, orange or white. Widely cultivated as an ornamental and available commercially in many colours. Ideal for sunny rock garden pockets, window-boxes and shallow containers, and suitable for inter-planting with low-growing fynbos bulbs like *Babiana angustifolia* and *Ixia maculata*. Sow the fine seed in autumn.

Liesl van der Walt

Liesl van der Walt

Felicia elongata

Family: Asteraceae

Height: 0.2–0.3 m

Flowering time: spring

A beautiful spreading annual with hairy lance-shaped leaves and yellow flower heads with white or mauve ray florets carried on long hairy peduncles. The ray florets are prominently banded with deep maroon at the base. Although rare in its natural habitat on the Cape West Coast, it responds well to cultivation and must have full sun to flower well. Prefers alkaline, sandy soil but can also be grown in slightly acidic loamy or granitic soils provided they are well-drained. Sow in autumn.

Felicia heterophylla

Family: Asteraceae

Common name: bloublomastertjie

Height: 0.2–0.35 m

Flowering time: spring

A very striking branched, low-growing, hairy-leafed annual with solitary flower heads borne on erect, hairy peduncles. The flower heads are dark blue with bright blue ray florets. An ideal species for interplanting with other low-growing fynbos annuals like *Dimorphotheca pluvialis* and *Ursinia cakilefolia*, or medium-sized fynbos bulbs like *Babiana angustifolia* and *Watsonia laccata*. The soil must drain well as the plants don't grow well in overly wet conditions. Sow in autumn.

Liesl van der Walt

Heliophila coronopifolia

Family: Brassicaceae

Common names: blue flax, showy sunflax, sporrie

Height: 0.4–0.6 m

Flowering time: spring

An erect, fine-leafed annual with bright sky-blue flowers and a white or greenish centre. A very showy species on its own, or inter-planted with other annuals, especially *Dimorphotheca pluvialis* (as seen above), or fynbos bulbs like *Watsonia borbonica* subsp. *ardernei*. It prefers sandy, well drained conditions and does not grow well in excessively moist conditions during the winter growing period. Sow in autumn.

Liesl van der Walt

Nemesia strumosa

Family: Scrophulariaceae

Common names: leeubekkie, rooileeubekkie

Height: 0.3–0.4 m

Flowering time: spring

An erect, sparsely branched annual having lance-shaped dark green leaves with toothed margins. It bears very showy large white, orange, red, yellow, pink or mauve flowers with white hairy throats variously marked with blue or brown. The plants like well drained sandy soil in full sun and are suited to rock garden pockets, containers and window boxes. This species seldom re-sows itself from year to year and it is essential to collect the seeds in early summer in order to maintain it in cultivation. Sow in autumn.

Senecio elegans

Family: Asteraceae

Common names: wild cineraria, veld cineraria, strandblommetjie

Height: 0.3–1 m

Flowering time: spring to early summer

A very showy annual with dense, glandular hairs covering the stems and fleshy, finely divided leaves. The flower heads are yellow with attractive purple or white ray florets produced in dense, flat-topped clusters. Flower heads with double ray florets are occasionally encountered. The plants are wind-tolerant and suitable for difficult, dry coastal gardens. Suitable for both full sun and lightly shaded positions, and can be successfully inter-planted with other fynbos annuals and bulbs like *Ursinia speciosa* and *Watsonia borbonica*. Sow in autumn.

Liesl van der Walt

Ursinia speciosa

Family: Asteraceae

Height: 0.3–0.4 m

Flowering time: spring

A very showy sprawling annual with delicate lacy foliage. Its ray florets have blunt tips and its involucral bracts are rounded and papery in the upper portion. The large solitary flower heads are carried on long wiry stalks and are glossy black with attractive orange, yellow or sometimes white ray florets. An ideal subject for mass spring displays and suitable for inter-planting with other fynbos annuals like *Senecio elegans,* and fynbos bulbs like *Watsonia borbonica* subsp. *ardernei*. Sow the papery seeds in autumn.

Background photograph: The blue flax *Heliophila coronopifolia* flowering *en masse* near Langebaan (see page 86)

These two pages kindly sponsored by Silverhill Seeds

Arctotis hirsuta
Felicia dubia
Ursinia cakilefolia
Senecio elegans
alis
Brenda Szabo

Below: The elandsvy *Carpobrotus quadrifidus* sprawling over a granite outcrop near Cape Columbine (see page 93)

Opposite: *Lampranthus amoenus* (see page 95)

7. GROWING FYNBOS MESEMBS

There are 660 species (including 525 endemics) in the Aizoaceae, making it the fourth largest family in the Cape Flora.

A quick guide to germinating mesemb seeds

- An analysis of recent germination results for the Aizoaceae showed that 43% of the species studied gave a significant response to smoke
- Use only mature, plump, fully formed seeds
- Pre-soak seeds in aqueous smoke extract or a commercial smoke seed primer for 24 hours or smoke seed trays after sowing
- Seeds should be given a light dusting with a fungicide dressing to prevent post-emergence seedling infection
- Sow seeds in a sandy, well-drained soil medium
- Incubate in full sun under autumn temperature conditions, e.g., alternating 10°C (16 hours night) x 20°C (8 hours day)

Ten of the Best Fynbos Mesembs

(see also Appendix 2, page 191)

Liesl van der Walt

Carpobrotus deliciosus

Common name: gaukum

Height: up to 0.25 m

Flowering time: winter to early summer, but mainly in spring

This very floriferous, vigorous succulent perennial produces long trailing stems and almost straight, bright green fleshy leaves. Its shiny, bright purple or occasionally pink or white flowers have conspicuous cream-coloured anthers. Its delicious ripe fruits are sweet and make excellent jam. It thrives in both sandy and loamy, well-drained soils, rapidly forming a thick mat and is highly recommended for stabilizing steep banks and for seaside gardens in full sun. It is very waterwise and easily propagated from cuttings taken in autumn or from division of rooted, trailing stems taken at any time of year. Sow seeds in late summer or early autumn.

Carpobrotus edulis subsp. *edulis*

Common names: Hottentots fig, sour fig, suurvy

Height: 0.15 m

Flowering time: spring to early summer

A robust, succulent perennial with trailing stems up to 2 m long. It has fleshy erect or slightly curved dark green leaves and large yellow flowers maturing to pink with age. The ripe fruits are fragrant, yellowish in colour and edible. An excellent mesemb for stabilizing steep banks and sand dunes and grows in almost any well-drained soil. It is highly recommended for difficult seaside gardens in full sun and the flowers attract beetles and bees to the garden. Very waterwise and easily propagated from cuttings taken in autumn or from division of rooted, trailing stems taken at any time of year. Sow seeds in late summer or early autumn.

Carpobrotus quadrifidus

Common name: elandsvy

Height: up to 0.3 m

Flowering time: spring to early summer

A robust succulent perennial with trailing stems up to 2.5 m long, large, straight, greyish-green fleshy leaves and spectacular bright purple or occasionally pink or white flowers with bright yellow anthers. It closely resembles *Carpobrotus acinaciformis*, differing mainly in its much larger flowers and straight leaves. It is an excellent groundcover and thrives in both sandy and well-drained, loamy soils. The plants require full sun, are very waterwise and require a dry summer for best flowering results, failing which excessive foliage is produced. Highly recommended for large rock gardens and for stabilizing steep banks. It is easily propagated from cuttings taken in autumn or from division of rooted, trailing stems taken at any time of year. Sow seeds in late summer or early autumn.

Drosanthemum floribundum

Height: 0.03–0.15 m

Flowering time: spring to early summer

A dense, low-growing groundcover or shrublet with very short, fleshy leaves and an abundance of eye-catching bright mauve flowers with cream-coloured anthers. The plants like sandy, well-drained soil in full sun and are ideal for rock garden pockets and window boxes, and for trailing over hot, low garden walls. It is very waterwise and prefers a dry summer, failing which excessive leaf growth and minimal flowering occurs. Easily propagated from cuttings taken in autumn or from division of rooted trailing stems taken at any time of year. Sow seeds in late summer or early autumn.

Drosanthemum speciosum

Common names: scarlet dewflower, rooi douvygie

Height: up to 0.6 m

Flowering time: from late autumn to late spring

A much-branched, neat shrublet with short, sausage-shaped leaves covered with small bladder-like, glistening cells. The spectacular bright red to orange flowers have whitish centres and are borne singly on long wiry peduncles. An excellent rock garden plant or subject for a large terracotta container, requiring full sun and well-drained, loamy or granitic soil. It prefers a dry summer and is very waterwise. Easily propagated from cuttings taken in autumn. Sow seeds in late summer or early autumn.

Jordaaniella dubia

Common names: mat vygie, strandvygie

Height: up to 0.04 m

Flowering time: late winter to spring

A prostrate succulent with long trailing stems forming an attractive mat-like groundcover. These waterwise plants have erect, fleshy, narrow leaves and beautiful large golden-yellow, or white flowers tipped with pale pink. They prefer sandy soil but easily adapt to heavier soil types provided they are well-drained. Full sun and a dry summer period are required for successful flowering, failing which excessive leaf growth is produced. Highly recommended for rock garden pockets and difficult coastal gardens. It is easily propagated from cuttings taken in autumn, or from division of rooted trailing stems taken at any time of year. Sow seeds in late summer or early autumn.

Ernst van Jaarsveld

Lampranthus amoenus

Height: 0.3–0.45 m

Flowering time: late winter to early summer

A densely branched, neat small shrub with greenish-grey foliage and a profusion of bright purplish-magenta or white flowers with yellow anthers. This long-lived, easily grown species is ideally suited to mass planting on well-drained slopes, or in rock garden pockets, pavement beds or large containers, in full sun positions. It prefers a dry summer and thrives in both acid and alkaline soils and tolerates high winter rainfall, but is also very waterwise. Easily propagated from cuttings taken in autumn, or from seeds sown in late summer or early autumn.

Lampranthus aureus

Common name: vygie

Height: 0.2–0.4 m

Flowering time: early spring

A low growing, small sub-shrub with attractive dark green, long fleshy leaves and spectacular large, bright orange flowers in spring. Although very rare in its natural habitat along the Cape west coast, it responds extremely well to cultivation and is an ideal subject for growing in pots, window boxes and rock gardens, or as an edging plant to larger perennials, in full sun. The plants prefer a dry summer and perform best in well drained granitic or loamy soils but can also be grown successfully in sandy media. Easily propagated from cuttings taken in autumn, or from seeds sown in late summer or early autumn.

Ernst van Jaarsveld

Lampranthus multiradiatus

Height: 0.3–0.6 m

Flowering time: late winter to early summer

A very floriferous low-growing, spreading shrublet with attractive fleshy grey foliage, densely covered with shell pink or dark red flowers. It is best planted in large groups for maximum effect and makes an excellent subject for rock gardens, window-boxes, large terracotta pots or for growing over low garden walls. It prefers a dry summer and thrives in poor sandy soil but easily adapts to heavier loam media, provided they are well-drained. Full sun, moderate watering in winter and a dry summer period are required for best results. Easily propagated form cuttings taken in autumn, or from seeds sown in late summer or early autumn.

Lampranthus roseus

Height: 0.3–0.5 m

Flowering time: spring to early summer

A compact, low spreading perennial with relatively long dark green leaves and masses of pale pink to rose-pink flowers with bright yellow anthers. It is highly recommended for mass planting on well drained slopes or in rock garden pockets, pavement beds and window-boxes, in full sun positions. It prefers a dry summer, performs equally well in alkaline and acid soils and is very waterwise once fully established. Easily propagated from cuttings taken in autumn, or from seeds sown in late summer or early autumn.

Opposite: *Carpobrotus quadrifidus* in habitat near Cape Columbine (see page 93)

Below: The endangered *Geissorhiza darlingensis* under cultivation in the Kirstenbosch bulb nursery (see page 191)

Opposite: *Moraea insolens*, an endangered species from the Caledon district (see page 192)

8. GROWING FYNBOS BULBS

The Cape Floristic Region, a relatively narrow landmass along the western and southern Atlantic coastlines of South Africa, is home to the world's richest and most beautiful geophyte flora, popularly known as bulbous plants. Nowhere else on earth has a more diverse and colourful bulbous flora evolved than here at the southern tip of Africa, comprising a staggering total of well over 1000 species. Stretching from just north of Nieuwoudtville in the Northern Cape, to Port Elizabeth in the Eastern Cape, the dominant fynbos vegetation type supports bulbous species on two major soil types, mineral-poor, acid sandstones and fertile, heavy clays. A substantial number of bulbous species from both these soil types is eminently suited to cultivation, almost all of which are waterwise to a greater or lesser degree. In addition, a large number has become vulnerable or endangered in their natural habitats, due mainly to burgeoning urban sprawl, alien plant infestation, industrial expansion and agricultural extension. By cultivating fynbos bulbs, an important contribution is made towards saving many of these species from extinction.

The popular term 'bulbous' usually refers to plants that survive unfavourable conditions (such as drought) by means of buds attached to specialized, subterranean storage organs. These include true bulbs (e.g. *Amaryllis, Lachenalia*), corms (e.g. *Gladiolus, Watsonia*), tuberous rootstocks (e.g. *Bulbinella*, *Zantedeschia*) and rhizomatous rootstocks (e.g. *Agapanthus, Kniphofia*). The bulbous flora occurring in fynbos can conveniently be placed into two major groups: winter-growing species, and evergreen species, of which the winter-growers far outnumber the evergreens. The growth cycle of most members of the winter-growing group is characterized by the production of new vegetative growth in autumn, as soon as temperatures begin to fall markedly after the long, hot, dry summer. This is followed by rapid vegetative growth during winter, and flowering from late winter to early summer, with a peak flowering period in spring. The growth cycle of most members of the family Amaryllidaceae is somewhat different in that flowering takes place in late summer or autumn, before vegetative growth begins. The small group of evergreen species flower mainly in summer, and also undergo a short dormant period in winter while maintaining their older leaves, generally producing new foliage in spring and summer.

Below: *Brunsvigia orientalis* in coastal fynbos at the Cape of Good Hope Nature Reserve (see page 191)

Opposite: The peacock flower *Spiloxene capensis* (see page 192)

Ernst van Jaarsveld

Seed production and dispersal takes place mainly in early summer for the deciduous winter-growers, and in late summer for the evergreens.

Growing bulbs in containers

With the obvious exception of the more robust fynbos bulbs such as the larger watsonias, many fynbos species can be very successfully cultivated in containers. In many instances, container cultivation is the only practical manner in which to grow the more delicate or fastidious fynbos members of *Gladiolus* and *Ixia*, for example.

Aspect and growing medium

A sunny aspect with free air circulation is required for all fynbos bulbs. In areas with mild winters, pots can be arranged in groups on a stoep or patio, and flat-dwellers can use window-boxes on a sunny balcony. It is important that the pots are not placed where they will overheat on very hot days. In areas with heavy winter rainfall, such as in the southern suburbs of the Cape Peninsula, the more delicate species are best grown under cover. In summer rainfall areas, fynbos bulbs are best grown under cover to provide protection from moisture during the summer dormant period. Fynbos bulbs are sensitive to frost, and generally do not perform well in humid environments.

Perfect drainage of the growing medium is one of the most important factors when cultivating most fynbos bulbs. Although many of them occur in nutrient-poor, well-drained sandy soils as well as in richer, heavier clay soils in the wild, under cultivation most of them are best grown in rather poor, well-drained growing media. Avoid the temptation to grow them in rich, water-retentive soil.

The most important component is sand, which should be a medium-grained, washed river or industrial sand. For easily cultivated species like *Gladiolus carneus* and *Moraea aristata*, a medium consisting of two parts river or industrial sand, one part loam and one part fine compost is recommended. For less easily cultivated species such as *Babiana rubrocyanea* and *Gladiolus trichonemifolius*, the amount of loam should be reduced considerably, or dispensed with entirely. Difficult species such as *Gladiolus debilis* and certain amaryllid genera such as *Cyrtanthus* should be grown in a medium of three parts river or industrial sand and one part fine compost; or the compost can be dispensed with entirely and a mixture of equal parts river and industrial sand used. Growers will discover their own ideal growing medium, but there can be no doubt that the more sand incorporated into the growing medium, the better the results will be.

Above: The critically endangered *Moraea gigandra* performs extremely well under cultivation (see page 192)

Opposite: *Kniphofia praecox* (see page 192)

A layer of broken crocks, stone or bark chips should always be placed over the drainage holes at the bottom of the container, and a 3 cm layer of fine compost should be placed over this, into which the roots can grow. This also prevents the finer growing medium from washing out following watering. Fill the rest of the container with the appropriate growing medium. Ordinary deep, brown plastic pots are ideal - a 15 cm pot is suitable for low-growing species like *Lachenalia contaminata* and *Oxalis hirta*, while a 20 cm pot is suitable for medium-sized species of *Geissorhiza* and *Romulea*. Taller species such as *Gladiolus carneus* and *Ixia viridiflora* require a 25 cm pot. A 30 cm pot is recommended for those species with vigorous root systems such as *Moraea aristata* and *Watsonia humilis,* while a 35 cm pot is recommended for species with very large bulbs, like *Brunsvigia orientalis*.

Planting

Depending on the species, the rootstocks are planted out from late January to mid-May, April being the most suitable month for most fynbos bulbs. Early-flowering bulbs like *Amaryllis belladonna* should be planted in late January or early February, just before the flower buds emerge, while the corms of *Moraea* are best planted in late autumn. Fynbos members of the family Amaryllidaceae, most of which have perennial fleshy roots, should not be disturbed once they are established, but if they have to be transplanted, it should be done immediately after the new leaves start to appear, while the bulbs are in active growth.

The depth of planting depends on the species but, as a general rule, they should be planted at a depth of about three times the height of the rootstock. Exceptions to this rule include the genera *Babiana* and *Cyanella*, which are planted twice as deep, while ornithogalums and most amaryllids are planted with the top of the neck at, or just below, the surface.

Watering

Once planted, pots should be watered well, and then not again until the leaf shoots begin to appear, after which a good soaking every fortnight is recommended for most species, depending on weather conditions. Over-watering of container-grown bulbs soon leads to rotting, and as a general rule it is preferable for the growing medium to dry out substantially between watering, rather than remain constantly wet. This applies particularly to amaryllids like *Brunsvigia* and *Nerine.* Exceptions to the general rule are certain species of *Geissorhiza* like such as *G. mathewsii* and *G. radians*, as well as *Onixotis stricta (=O. triquetra)* which require a continually moist medium during the winter growing period. Another exception is the evergreen *Wachendorfia thyrsiflora* that requires poorly drained, boggy conditions throughout the year in order to flower well.

Towards the end of spring, as temperatures rise, the plants begin to go dormant, which is indicated by yellowing of the leaves. Watering must then be withheld completely, and as soon as seed has been harvested and the foliage has completely withered, containers can be placed in a cool dry place and stored for the summer. Dormant bulbs of deciduous species grown in the garden should also be kept as dry as possible in summer.

Growing bulbs in the garden

While many fynbos bulbs are recommended for container cultivation, most are unsuited to general garden culture due mainly to their delicate nature, extremely short flowering period, inability to withstand garden irrigation during their summer dormant period and the exploits of moles and porcupines. (See 'Ten of the best fynbos bulbs' at the end of the chapter for a list of species recommended for garden cultivation).

Aspect and growing medium

As with container subjects, a sunny aspect with free air circulation is required for fynbos bulbs when grown in the garden. Soil must be very well drained, but generally species suited to garden culture are able to withstand less well-drained soils than those that can only be grown in containers. Drainage can be improved by mixing in large quantities of well-decomposed compost and river sand. Slightly sloping ground is ideal for planting as it allows for good water run-off. The rockery is a suitable spot in which to display groups of fynbos bulbs of the same species, but where moles are prevalent, the smaller species have to be grown in plunged pots covered with chicken wire. Fynbos bulbs are displayed to great advantage by inter-planting with low-growing spring annuals such as *Dorotheanthus bellidiformis* and *Nemesia strumosa.*

Planting

Depending on the species, the rootstocks are set out from late January to mid-May, at the same depths recommended for container subjects, but in extremely sandy soil they can be planted deeper.

Watering

After planting, the rootstocks should be watered well and not again until the leaf shoots appear, after which a fortnightly soaking can be given, if natural precipitation is lacking.

The species recommended for garden culture are generally those that can withstand a fair amount of garden irrigation during the dormant period, but if one is unable to lift, store and re-plant them every year, they are best planted in areas of the garden that receive as little water as possible during summer.

Propagation and care of fynbos bulbs

Seed

The seeds of winter-growing and evergreen fynbos species are sown in autumn. Exceptions are the seeds of *Agapanthus africanus, Agapanthus praecox, Cyrtanthus* species and the fleshy seeds of all other members of the family Amaryllidaceae, which should be sown as soon as they ripen from late summer to late autumn. Generally speaking, fresh dormant seeds of fynbos bulbs germinate readily without the need for smoke treatment.

Below: *Romulea obscura* (see page 192)

Deep seed trays or pots should be used and the sowing medium should preferably be sterilized beforehand with boiling water in order to kill weed seeds. The sowing medium used will depend on the species, but a good general medium is equal parts river or industrial sand, and fine compost or loam. For the more delicate species, the amount of compost or loam should be reduced. Seeds should be sown thinly to prevent overcrowding and allow sufficient room for the developing rootstocks. The seeds of most species need only be covered with a thin layer of sand (3–4 mm), while the large fleshy seeds of amaryllids are simply pressed into the medium to rest at, or just below, soil level. An exception is the genus *Cyrtanthus* whose flat, light dry seeds can also be germinated by placing them in glass containers filled with water. The water should be replaced once a week and when the seedlings have produced a few leaves, transfer them to pots or seed trays. The seedlings of all fynbos species should remain in the seed tray or pot for at least one full season; in many instances they should remain undisturbed for 2 to 3 seasons before being planted out into permanent containers or into the garden.

Offsets, bulbils and cormels

Offsets formed on fynbos bulbs and corms are removed during the dormant period, when large enough. Division of corm offsets is achieved simply by breaking up the corm clusters by removing and discarding old corms. Corm offsets can be stored dry until the following planting time, but bulb offsets of species with perennial fleshy roots such as *Nerine sarniensis*, should be re-planted immediately. Several species of *Lachenalia* reproduce by bulbil formation on leaf bases, or at the tips of subterranean stolons, such as *L. bulbifera*. Similarly, several members of the Iridaceae produce cormels at the tips of stolons, such as *Ixia maculata*. Bulbils and cormels are removed during the dormant period and stored until planting time in autumn.

Division of rhizomatous rootstocks

Fynbos species with rhizomatous rootstocks such as *Aristea capitata* and *Kniphofia praecox* are propagated vegetatively by lifting the clumps and prising them apart with 2 forks placed back to back in the centre. The foliage is then cut back by about one third and the individual portions of rootstock are re-planted as soon as possible.

Leaf cuttings

Propagation by leaf cuttings is an effective way of increasing stocks of fynbos *Lachenalia* species. Leaves for cutting material should be virus-free and preferably in active growth. Depending on leaf size, the leaf material is cut into cross sections and placed in a well-drained rooting medium such as equal parts river sand and vermiculite, with the base of the cutting about 1 cm below the surface. The cuttings are placed in a shaded position and kept only slightly moist. Bulblets begin to form at the base of the cutting after about one month. Remove and store at the end of the growing season and plant out in autumn.

Feeding

Fynbos bulbs can, in general, be grown successfully without any supplementary feeding because of their low nutritional requirements, but this is not to say that feeding is not recommended. With the exception of amaryllids like *Amaryllis* and *Nerine* that should not receive any feeding, most fynbos species respond very favourably to fertilizers with a high potash but low nitrogen content. Slow-release products such as the non-toxic, non-burning fertilizer Neutrog Bounce Back can be incorporated into the upper part of the growing medium or sprinkled on the surface. It provides water insoluble nitrogen that is released over a long period, and supplies a full balance of nutrients that do not leach rapidly. Liquid fertilizers, such as Kelpak 66, can be used at a weaker rate than recommended, at fortnightly intervals.

Pests and diseases

Under cultivation, fynbos bulbs are at times subject to various pests and diseases.

Pests

Aphids are small green or black sucking insects usually found on developing flower buds of irids like *Babiana*, *Freesia*, *Moraea* and *Watsonia*. Make up a solution of 5 ml liquid soap in 1 ℓ of water. Using a small spraying can, spray the aphids away with the solution. Ladybirds are natural predators of aphids. Alternatively, in heavy infestations, use mineral oil (e.g. Oleum) as a full cover spray.

Lily borers (amaryllis caterpillars) are highly destructive black and yellow striped caterpillars that only attack amaryllids, rapidly boring into leaf, flower and stem tissue, causing it to turn black and disintegrate. Remove affected parts as soon as noticed, and crush by hand or under foot. Alternatively, use carbaryl (e.g. Karbaspray) as a full cover spray.

Mealybugs are small, white, waxy sucking insects found in large numbers between bulb scales and corm tunics, as well as at the bases of leaves. They thrive under enclosed, warm conditions and are especially partial to container-grown bulbs. They secrete honeydew, are spread by Argentine ants and are transmitters of viral disease. Remove them with tweezers or squash them against the bulb or corm. Alternatively, mix equal parts of methylated spirits and water, then dip a cotton wool bud into the solution and wipe the insects away. In severe infestations of container-grown bulbs, drench the soil with mineral oil (e.g. *Oleum*).

Mound-forming molerats are especially fond of fynbos species with true corms such as babianas, freesias, gladioli, moraeas and dwarf watsonias. As a measure of control, plant the corms in plastic pots, secure the upper surface with chicken wire by tying it down with a single strand of strong wire placed under the rim of the pot, and plunge to just below soil level until flowering and seed formation have been completed.

Protecting fynbos bulbs from **porcupines** is almost an impossible task! As a suggested measure of control, maintain a stock of their favourite meal of *Zantedeschia aethiopica* rootstocks some distance away from the area chosen for one's more treasured fynbos bulbs. Alternatively, scatter the pungent scatter crystals known in the trade as 'Get Off My Garden' that are meant to confuse dogs' and cats' sense of smell.

Red spider mites are tiny red, spider-like mites that are most prevalent as temperatures rise towards the end of spring and throughout summer. They attack the foliage of many irids, especially babianas and gladioli, and form a dense, spider-like web over the leaf surfaces.

Watsonia borbonica subsp. *borbonica* (see page 115)

Apply mineral oil (e.g. Oleum) as a full cover spray.

Slugs and snails are active by night and damage the leaves of numerous fynbos bulbs, especially *Amaryllis, Brunsvigia, Lachenalia, Moraea* and *Sparaxis*. Sprinkle salt directly onto active slugs and snails, or apply tobacco dust in a circle a short distance away from the base of the plant. In heavy infestations, apply metaldehyde (e.g. Snailban) as a bait.

Snout beetles are small grey, nocturnal beetles that cause tremendous damage to the foliage and flower stems of amaryllids like *Nerine*, and hide between the leaf bases by day. Using a bright torch, pick off the beetles at night by placing a dish under affected leaves, shaking them off and crushing them by hand. In severe infestations, apply cypermethrin (e.g. Garden Ripcord) as a full cover spray.

Diseases

Damping off fungi are prevalent under inadequately aerated conditions, especially where seeds have been sown too thickly. Apply captab (e.g. Kaptan) as a preventative seed-dusting treatment prior to sowing, or as a soil drench.

Fungal rotting of rootstocks should be suspected when leaves are slow to develop, look unhealthy, or fail to appear at all. Lift and thoroughly clean the rootstock in water, cut away all infected parts, dust completely with captab (e.g. Kaptan), and re-plant in slightly damp river-sand to encourage new root growth.

Rust attack of the foliage of fynbos bulbs is most prevalent in *Ixia* and *Watsonia*, and is seen as reddish-brown pustules that break open and liberate powder-like spores. In severe infestations apply mancozeb (e.g. Dithane) preventatively as a full cover spray.

Viral symptoms are recognized as discolouration patterns of leaves and flowers in the form of mosaics, mottling or light-coloured streaks. Such plants should be isolated as soon as possible and destroyed, as no cure exists.

The above mentioned chemicals are poisonous and potentially dangerous and should be applied with great care.

Below: *Moraea tulbaghensis* is an endangered but easily cultivated plant (see page 192)

Opposite above: *Lachenalia aloides* var. *aurea* makes an outstanding container subject (see page 113)

Opposite below: *Ixia dubia* (dark orange form) (see page 191)

Above: *Ixia scillaris* flowering *en masse* after a fire in the Biedouw Valley (see page 192)

Below: *Ammocharis longifolia* in habitat near Saldanha (see page 191)

Ten of the Best Fynbos Bulbs

(see also Appendix 2, pages 191–193)

Amaryllis belladonna

Family: Amaryllidaceae

Common names: March lily; belladonna lily

Height: 0.75–0.90 m

Flowering time: late summer to autumn

The showy, sweet-scented, bell-like flowers appear in late summer and early autumn from bone-dry earth, and are followed by strap-shaped leaves that grow from late autumn to early summer. Plant the large egg-shaped bulbs with the top of the neck at, or just above soil level, in well-drained, sandy soil. The bulbs should not be planted in rich soils and require no feeding. Once established, they like to remain undisturbed for many years until overcrowded, and flowering performance diminishes. The leaves must have access to bright light or direct sun in order to flower, however flowering is erratic and not every bulb flowers every year. The bulbs are ideal for rockeries, are waterwise and withstand irrigation during their summer dormant period. It is highly susceptible to lily borer (amaryllis caterpillars). Sow the fleshy seeds as soon as they ripen in late autumn.

Babiana pygmaea

Family: Iridaceae

Common name: geelbobbejaantjie

Height: 0.05–0.2 m

Flowering time: late winter to early spring

A beautiful dwarf species with large cup-shaped yellow flowers and purplish centres, and strongly ribbed, hairy lance-shaped leaves. It is endangered in its natural habitat in the southwestern Cape but is easily cultivated in well-drained, sandy soil in sunny positions. It is best grown as a subject for deep containers that can be displayed on a sunny verandah or plunged into rock garden pockets for the flowering period. The corms are sensitive to summer moisture and must be lifted and stored completely dry during this period. Easily propagated by corm offsets or by seeds sown in autumn. Seedlings will flower in their second season under ideal conditions.

Chasmanthe floribunda

Family: Iridaceae

Common names: suurkanol, Adam's rib

Height: 0.5–1.2 m

Flowering time: early winter to early spring

One of the most easily grown fynbos geophytes with attractive spikes of tubular orange or yellow flowers that attract sunbirds and white-eyes. The large, flat corms multiply rapidly and require replanting every few years in order to flower well. It flourishes in light shade and full sun, and is waterwise. The dead leaves should be cut down in early summer. It tolerates coastal conditions and grows in almost any soil. The primrose-yellow variety, *Chasmanthe floribunda* var. *duckittii* (above) is highly recommended and commences flowering in early winter, slightly earlier than the orange-flowered var. *floribunda*. Both varieties are useful in complementary plantings with evergreen or winter-deciduous *Agapanthus* species, such as *A. praecox* subsp. *minimus* 'Adelaide' and *A. inapertus* subsp. *pendulus* 'Graskop', respectively, and easily withstand general garden irrigation over the summer dormant period. Sow the seeds in autumn.

Freesia alba

Family: Iridaceae

Common names: freesia, ruikpypie

Height: 0.12–0.4 m

Flowering time: late winter to late spring

The long-lasting, heavily scented flowers are pure white or may have a pale to deep yellow marking on the lower tepal, and are borne on sturdy, branched inflorescences. The corms need moist conditions throughout winter and are completely dormant over summer, during which time they should preferably be kept dry, but will usually survive garden irrigation if the soil is sufficiently well drained. They are most successfully grown as specimen plants in containers, displayed on a sunny patio or plunged into the garden in rock garden pockets, then lifted and stored dry for the summer. They make excellent, long-lasting cut-flowers and are easily raised from seeds sown in autumn, or by separation of cormlets during the summer dormant period.

Gladiolus carneus

Family: Iridaceae

Common names: painted lady, bergpypie

Height: 0.2–0.6 m

Flowering time: spring to summer

A very attractive, floriferous white or pink-flowered species with prominent dark pink markings on the lower tepals, and sword-shaped foliage. It occurs in the south-western and southern Cape and its numerous long-lasting blooms are usually borne on a branched inflorescence and make excellent cut-flowers. It requires full morning sun or bright light for as much of the day as possible, and can be grown as a specimen container subject or plunged into sunken wire baskets in rock garden pockets. The corms must be kept dry during the summer dormant period and it is easily propagated from offsets, or seeds sown in autumn that will flower in their second year under ideal conditions.

Lachenalia aloides

Family: Hyacinthaceae

Common names: geelklipkalossie, geelviooltjie

Height: 0.15–0.3 m

Flowering time: late winter to spring

A very rewarding, deciduous, winter-growing bulb with drooping, tubular flowers (see the golden-yellow *L. aloides* var. *aurea* on page 109) and spotted fleshy leaves. It is most suitably grown as a container subject under cover in bright light, or receiving morning sun and afternoon shade. The blooms are particularly long lasting (up to three weeks) and suitable as cut-flowers for small arrangements. The bulbs multiply rapidly and like a sandy soil mixed with some finely sifted compost, regular watering in winter and spring, and a completely dry summer rest. It can also be propagated from seeds sown in autumn and will flower in its third year. The particularly showy *L. aloides* var. *quadricolor* (above) flowers in winter and multiplies rapidly by offsets.

Moraea elegans

Family: Iridaceae

Common name: poutulp

Height: 0.25–0.4 m

Flowering time: early to late spring

A striking and variable yellow-flowered plant with prominent green, orange or brown markings on the outer tepals. It displays a succession of flowers over a six-week period and produces a single, long, narrow leaf from a small corm that requires moisture from autumn to late spring, but must be kept absolutely dry over the summer dormant period. It requires regular watering in winter and spring and is ideal for full sun positions in well-drained rock gardens and deep containers, or placed towards the centre of a fynbos herbaceous border. The corms readily produce offsets, and plants are easily raised from seeds sown in autumn that will flower in their second year, under ideal conditions.

Veltheimia bracteata

Family: Hyacinthaceae

Common names: forest lily, sandui

Height: 0.3–0.6 m

Flowering time: midwinter to late spring

The imposing, long-lasting poker-like inflorescences appear amid a rosette of shiny, dark green leaves with undulate margins. The foliage of the pink forms remains almost evergreen in temperate climates, producing a new rosette in early autumn. It prefers well-composted, well drained soil but thrives in almost any growing medium and is ideally suited to lightly shaded, difficult corners of the garden, or planted in containers or large drifts under evergreen or deciduous trees. The striking greenish-yellow cultivar *Veltheimia bracteata* 'Lemon Flame' prefers a sunnier position and usually maintains a distinct summer dormant period. Easily propagated by bulb offsets taken in late summer, or by seeds sown in autumn.

Watsonia borbonica subsp. *borbonica*

Family: Iridaceae

Common name: suurkanol

Height: 1–2 m

Flowering time: early to midsummer

A very easily cultivated, winter-growing plant with sturdy, tall branched spikes of deep pink, bell-shaped flowers. It must have full sun and is suited to inter-planting with spring-flowering annuals in rockeries, as a massed display, or as clumps in a mixed fynbos garden. The corms are planted in autumn and can remain undisturbed for several years until overcrowding necessitates corm separation. The equally showy, white *Watsonia borbonica* subsp. *ardernei* (see page 184) flowers slightly earlier and is grown in exactly the same manner. This choice garden subject is also a desirable cut-flower. The corms multiply rapidly and are best kept as dry as possible over the summer dormant period. Sow seeds in autumn.

Watsonia laccata

Family: Iridaceae

Common name: rooikanolpypie

Height: 0.2–0.4 m

Flowering time: spring to early summer

A very floriferous dwarf species with orange, pale to deep pink or white blooms and short, lance-shaped leaves. It likes a sunny position and is most successfully grown as a container plant, or plunged into rock garden pockets in autumn and lifted in early summer. The corms multiply rapidly but are very sensitive to summer moisture and have to be kept completely dry over this period. It is an ideal subject for mixed plantings with low-growing fynbos annuals like *Dorotheanthus bellidiformis* and *Felicia heterophylla*. Easily raised from seeds sown in autumn, it will flower in the second season under ideal conditions.

Background photograph: Deep orange form of *Ixia dubia* (see page 191)

 These two pages kindly sponsored by Silverhill Seeds

Chasmanthe bicolor
Freesia alba
Moraea gigandra
Veltheimia bracteata
neus

Below: *Agathosma ovata* 'Kluitjieskraal' is one of the most outstanding buchus to grow (see page 127)

Opposite: *Adenandra fragrans* has highly aromatic foliage (see page 125)

9. GROWING FYNBOS BUCHUS

(With contributions from text by Fiona Powrie and Alice Notten)

The word 'buchu' is the Khoikhoi word for aromatic herb and is commonly used to describe the group of small-leafed indigenous shrubs that exude a pleasant, fresh smell when touched.

The Rutaceae (the economically important cosmopolitan family that includes citrus fruits) to which the buchus belong, is the ninth largest family in the Cape Flora with 273 species and 258 endemics. Many of the species are of importance as sources of volatile oils for the industries producing perfumes, cosmetics, soap and food colourants.

The genera included under the term buchu are *Acmadenia, Adenandra, Agathosma, Coleonema, Diosma, Empleurum, Euchaetis, Macrostylis, Phyllosma* and *Sheilanthera.*

The common characteristic is the presence of volatile oils in glands found on the leaves and fruits. The commercially grown 'true buchus' are *Agathosma crenulata* and *Agathosma betulina.* Oil is extracted for use in manufacturing cosmetics, soaps and food colourants and medicinally for

Agathosma glabrata (right foreground) in an indigenous garden setting (see page 127)

the treatment of renal disorders and chest complaints.

Buchu plants make a valuable contribution to any garden. They are small to medium shrubs with a neat compact appearance, providing colour in the winter with a proliferation of flowers ranging from white to pink and lilac, and delightful aromatic foliage. The flowers are small (5–20 mm in diam.), with 5 petals arranged in a star shape, and occur either singly or clustered together to resemble a small pompon.

Cultivation

The main distribution range of buchus is in the southern and south-western Cape. As minor members of the fynbos vegetation, they have the fynbos fire-prone life span of between 10 and 15 years, becoming more woody as they age. They grow best when planted in full sun. The soil should be slightly acid, well drained, composted and enriched with a well-balanced fertilizer.

Buchus should be planted out in groups of 3 to 5 with a spacing of 200–300 mm between plants. They respond to fairly dense planting which helps to retain soil moisture. An annual mulching of well-rotted compost is recommended to suppress weeds, keep soil temperatures low and reduce loss of moisture. Buchu plants always occur naturally in a mixture of fynbos plants and ideally should be used in this way in a garden; with restios, phylicas, pelargoniums, polygalas, and helichrysums as companion plants.

Young buchu plants should be pinched back to encourage bushy growth. Cultivating soil around the root area is not advised. Water requirements are moderate and when plants are irrigated it is advisable to have the foliage dry by nightfall to discourage any fungal attack.

Pests and diseases

The volatile oils in buchus, responsible for their aromatic and sometimes pungent smell, act as an anti-feedant to discourage insect attack. The exception is the citrus caterpillar that can be controlled with the use of a contact stomach insecticide. *Phytophthora cinnamomi,* a soil and water-borne fungus, attacks many fynbos species. Growth of this fungus is promoted by high soil temperatures and it attacks the plant's root system, preventing the uptake of water. The plants wither rapidly and die. Treat young plants with Furalaxyl (acylalanine) or Fongarid 25 WP just prior to planting out. Take care to minimize root damage. Use a soil mulch to reduce soil temperature. Disturb the soil as little as possible. Remove infected plants immediately and drench space with diluted Jeyes Fluid. This is not a cure-all, but a measure of control to protect adjacent plantings.

Propagation by seed (see also page 123)

The buchus produce their seeds in capsules from which they are expelled on ripening. This phenomenon is known as ballistic dispersal. To harvest the seed, wait for the first few capsules to open and then collect the plump darker capsules. Place in a closed paper bag or lidded box (not plastic). Store in a warm, dry place and the seed will be expelled. In fact, you will hear it happen. Sowing is usually done in April.

Prepare a well-drained medium using sand, loam and compost in equal proportions. The mix should be light. Place the medium in l00 mm deep trays with adequate drainage holes. Level and firm the surface and water well. Sow seed evenly and cover with approximately 3 mm of medium. Apply a smoke treatment (see page 181) as this will greatly enhance germination. Alternatively, seeds may be soaked for 24 hours in a solution of commercial seed primer before being sown. Seed trays should be placed in a covered area with good light and air circulation. Seed trays should be kept damp. Over-watering and poor drainage will favour bacterial infection. Seeds take 1 to 2 months to germinate. The germination percentage is often extremely low. An analysis of recent germination results for the Rutaceae showed that 20% of the species tested gave a significant response to smoke treatment.

Agathosma glabrata 'Purple Sunrise' is an extremely floriferous plant (see page 127)

Liesl van der Walt

When 4 true leaves have developed, seedlings are large enough to be pricked out into 0.5 ℓ bags containing a fynbos potting medium. Great care should be taken not to damage the fine root system. Seedlings should be hardened-off in a protected area for 3 weeks and then placed in the sun. A regular application of a balanced liquid fertilizer is recommended. The growing tips of the seedlings should be pinched out to encourage bushy growth. The plants need about nine months to develop before being planted out in the garden. Flowers are produced after 2 years.

A quick guide to germinating buchu seeds

- Use only mature, fully-formed seeds
- The seeds should be pre-treated by placing them in hot water at 100°C to soften the woody seed coats. A light scarification treatment using fine sandpaper or a short soak in concentrated sulphuric acid is also sometimes used to soften the seed coats prior to soaking in seed primer.
- Pre-soak seeds in a commercial smoke seed primer for 24 hours
- Seeds should be given a light dusting with a fungicide dressing to prevent post -emergence seedling infection
- Sow seeds in a sandy, well-drained slightly acid soil medium
- Alternatively, seed trays may be smoked in a smoke tent once the seeds have been sown
- Incubate in full sun under autumn temperature conditions, e.g. alternating 10°C (16 hour night) with 20°C (8 hour day)

Vegetative propagation

Cuttings have the advantage of producing a larger flowering plant more quickly than seedlings. Semi-hardwood cuttings, 50–70 mm long, are taken from the current year's growth. Suitable material is produced 6 to 9 weeks after flowering, usually from August to early October. Prepare cuttings by making a clean cut below a node and remove the lower third of the foliage. Dip cut ends in a rooting hormone such as Seradix 3. Firmly place cuttings in a medium of 50% bark and 50% polystyrene. Ideally, these cutting trays should now be placed in a well-aerated propagation unit with a bottom heat of 24°C with intermittent misting. However, a measure of success can be obtained by enclosing the cutting tray in transparent plastic to create a terrarium. The plastic covering should not have any contact with the foliage and the rooting medium should always remain damp.

Rooting occurs in 9–11 weeks. Carefully pot the rooted cuttings into 0.5 ℓ bags. A well-drained humus-rich fynbos potting medium should be used. Plants should be hardened-off in a protected area for 3 weeks before being placed in the sun. Plants should be ready for planting out in the open within 7–8 months.

Plants should be fed regularly with a well-balanced fertilizer. Yellowing leaves can be treated with an application of iron chelate.

Ten of the Best Fynbos Buchus

(see also Appendix 2, page 194)

Liesl van der Walt

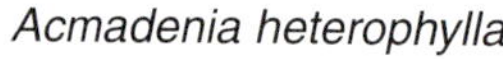

Acmadenia heterophylla

Height: up to 0.4 m

Flowering time: autumn

A low, spreading, compact perennial with aromatic fine leaves and solitary, small bright pink flowers. This is a floriferous, long-lived species, suitable for sunny rock garden pockets, as an edging plant to herbaceous borders, or as a filler in a mixed fynbos bed, and can also be successfully grown in deep containers. Highly recommended as one of the most attractive and easily grown buchus for both acid and alkaline soils, and ideally suited to inter-planting with dwarf fynbos bulbs like *Babiana angustifolia* and *Freesia alba*. Sow seeds in autumn.

Acmadenia mundiana

Height: 0.4–1 m

Flowering time: late winter to midsummer

An attractive small shrub with erect branches clothed in aromatic, lance-shaped leaves and terminal clusters of large, dusky pink flowers. It is suitable for mixed fynbos beds or grown *en masse*, or placed towards the front of a border. It performs equally well in sharply drained, alkaline and acid soils and is an ideal subject for inter-planting with medium-sized fynbos bulbs like *Moraea elegans* and *Watsonia laccata*. Light pruning of the branches after flowering maintains a compact growth habit and prevents plants from becoming too woody. Sow seeds in autumn.

Adenandra fragrans

Common names: anysboegoe, klipsissie

Height: 0.4–1 m

Flowering time: early spring to early summer

One of the most appealing of all the buchus with highly aromatic foliage and sticky terminal clusters of large pale pink flowers. It forms a small, compact shrublet and is suited to mixed fynbos beds and rock garden pockets in acid, well-drained soil, in full sun. Light pruning of the branches after flowering is essential to maintain a compact growth habit and prevent plants from becoming too woody. Suited to inter-planting with medium-sized fynbos ericas like *Erica lateralis* and *Erica regia*. Sow seeds in autumn.

Adenandra gummifera

Height: 0.3–1 m

Flowering time: early to late spring

A very striking buchu with oblong, aromatic leaves and large white flowers with a pink reverse, produced in very sticky terminal clusters. It is an easily grown species, well suited to mixed fynbos beds or rock garden pockets, and needs acid soil in full sun or lightly shaded positions. The branch tips should be pruned back after flowering to ensure a compact growth habit and prevent plants from becoming too woody. Ideal for inter-planting with fynbos ericas like *Erica cerinthoides* and *Erica regia*. Sow seeds in autumn.

Adenandra villosa

Height: 0.5–1 m

Flowering time: winter to early summer

A small shrub with aromatic elliptical leaves produced on reddish branches, and clusters of large, white starry flowers with pinkish-red undersides. It is an easily grown, long-lived plant, ideally suited to rock garden pockets, mixed fynbos gardens or as a container subject. It requires well-drained, acid soil in full sun positions. Light pruning of the branches after flowering ensures a compact growth form, and inter-planting with *Erica cerinthoides* or *Erica regia* creates a striking colour contrast. Sow seeds in autumn.

Agathosma crenulata

Common names: long-leaved buchu, asynboegoe, regteboegoe

Height: 0.5–3 m

Flowering time: winter to early summer

A robust, attractive rounded shrub with intensely aromatic, glossy, dark green, oval leaves that are very rich in oil glands. It is grown commercially for its oil and has been used medicinally for the treatment of renal and chest complaints. Large white to mauve flowers are borne in the axils of the leaves. It is a long-lived, easily grown species, ideally suited to mixed fynbos gardens. Light pruning of the branches after flowering is essential to maintain a compact growth habit and prevent plants from becoming too woody. It requires well-drained, acid soil and regular watering. Suited to inter-planting with larger ericas like *E. versicolor* and *E. verticillata.* Highly attractive to bees and butterflies. Sow seeds in autumn.

Agathosma glabrata

Height: 1–2 m

Flowering time: late winter to midsummer

A densely branched, strongly growing small shrub with aromatic lemon-scented foliage and masses of pink, mauve, purple or white flowers produced in terminal clusters. It is a long-lived, easily grown plant, highly recommended for mixed fynbos gardens, requiring well drained acid soil in full sun positions. A purple, low-growing form of this species known as *Agathosma glabrata* 'Purple Sunrise' (see page 122) is particularly attractive and suited to rock garden pockets or planted as a border, and is also ideal for inter-planting with medium-sized fynbos bulbs like *Gladiolus carneus* and *Watsonia laccata*. Highly attractive to bees and other insects. Sow seeds in autumn.

Agathosma ovata

Common names: false buchu, basterboegoe, bosboegoe

Height: 0.3–1 m

Flowering time: late winter to spring

A very variable, neat, rounded groundcover or small shrub with aromatic lance-shaped leaves and masses of small mauve, pink or white flowers produced in the leaf axils. It is a long-lived, easily grown species, ideal for the beginner, and is waterwise once well established. Numerous forms occur such as the pink-flowered *Agathosma ovata* 'Glentana', an upright, fast growing shrub, *Agathosma ovata* 'Igoda', another pink-flowered form with finely textured foliage, and *Agathosma ovata* 'Witteklip', a robust, white-flowered plant (see above). The well known *Agathosma ovata* 'Kluitjieskraal' (see page 118) is the most highly recommended, producing clusters of mauve or pink star-shaped flowers which cover the plant, creating a spectacular show in mixed borders, fynbos beds, pots and mass plantings. Sow seeds in autumn.

Coleonema aspalathoides

Height: 0. 4–1 m

Flowering time: autumn to summer, but mainly early spring

A densely branched, small shrub with slightly aromatic, fine linear foliage and solitary, showy bright pink flowers. It is an easily grown species requiring well drained, acid soils in full sun and is highly recommended for rock garden pockets or mixed fynbos beds (see page 195). It performs well in containers and is ideal for inter-planting with fynbos bulbs like *Babiana angustifolia* or ericas like *Erica baueri*. Light pruning of the branch tips after flowering keeps the plants neat and compact. Sow seeds in autumn.

Coleonema pulchellum

Common name: confetti bush

Height: 0.7–1 m

Flowering time: winter to spring

A neat, much-branched shrub with feathery, pungent, sweet-smelling foliage. Masses of miniature pinkish-white, star-like flowers cover the bush for many weeks and it is a long-lived, easily grown species in acid or alkaline soils. It is suitable for general garden planting or mixed fynbos beds or as a container plant, and can also be used as a low hedge or bonsai subject. This well known, waterwise buchu is recommended for inter-planting with tall-growing fynbos bulbs like *Aristea capitata* and *Watsonia borbonica*, and cut flowering stems can be used in a mixed fynbos display. Sow seeds in autumn.

Agathosma glabrata is an outstanding garden plant (see page 127)

Below: *Selago canescens* (pale mauve, centre and right foreground) is an excellent subject for mixed fynbos plantings (see page 135)

Opposite above: *Halleria lucida* (see page 135)

Opposite below: A fine specimen of *Halleria lucida* at Kirstenbosch (see page 135)

10. GROWING FYNBOS SCROPHULARIAS

There are 418 species, 297 of them endemic, in the Scrophulariaceae in the Cape Flora, making it the sixth largest family.

An analysis of recent germination results for the Scrophulariaceae showed that seed of the majority of species studied gave a significantly higher germination rate after smoke treatment. It is thus recommended that seed of all species of Scrophulariaceae be treated with smoke before sowing.

Ten of the Best Fynbos Scrophularias

(see also Appendix 2, page 194)

Freylinia densiflora

Common names: bell bush, klokkiesbos

Height: 2–3 m

Flowering time: midwinter to spring

An erect, much-branched evergreen shrub with short, lance-shaped dark green leaves and elegant clusters of tubular white, pale pink or mauve flowers. It is fairly waterwise once established and highly recommended for mixed fynbos gardens in full sun positions. It has a persistent rootstock that readily resprouts after the branches are cut back and is a useful plant for screening unsightly structures, or displayed towards the rear of garden beds. It is easily raised from seeds sown in spring or early summer and seeds germinate within three weeks. Soft to semi-mature, short stem cuttings can be taken in autumn, spring or summer, treated with a rooting hormone, will root within one month. Regular pruning after flowering encourages a more compact, tidy appearance.

Freylinia helmei

Height: 1–2.5 m

Flowering time: spring to early summer

An interesting, erect woody shrub with clusters of pendulous mauve and creamy-white tubular flowers and leathery, dark green elliptical foliage. It is a fairly vigorous species and an excellent choice for sunny fynbos gardens or placed towards the back of a border. It thrives in well drained loamy soil and once established is very waterwise. The plants have a persistent rootstock or lignotuber, and readily form new shoots from the base after pruning. Pruning of the branches after flowering is necessary to encourage a more compact, bushy growth habit. It is easily propagated from cuttings taken in autumn, spring or summer, treated with a rooting hormone, will root within one month. Sow seeds in spring or early summer.

Freylinia lanceolata

Common names: honey bell bush, heuningsklokkiesbos

Height: 2–6 m

Flowering time: sporadically throughout the year but mainly winter to early spring

An evergreen shrub or small tree with arching branches of willow-like foliage. Usually multi-stemmed but occasionally develops into a single-stemmed weeping tree. The golden-yellow, honey-scented tubular flowers are borne in attractive clusters and attract insects that provide nourishment for insect-eating birds. The plants are wind-resistant, and relatively frost hardy and pest-free. They prefer a sunny spot and grow well in both summer and winter rainfall areas, and can tolerate temperatures from -2°C–37°C. Pruning after flowering is necessary to keep them looking neat. Young plants are easily propagated from the tiny wingless seeds sown in spring or early summer. They can also be propagated from stem cuttings taken during the summer months and treated with a rooting hormone.

Freylinia undulata

Common name: mauve honey bell bush

Height: up to 2 m

Flowering time: winter to midsummer

An erect, densely branched, drought-resistant shrub with oval-shaped leaves and undulate margins. A profusion of white, blue, mauve or purple tubular flowers are borne at the tips of the branches in slender racemes and it thrives in both clay or loamy soils in full sun. It is easily raised from cuttings taken in autumn or spring, treated with a rooting hormone and placed in a rooting medium of equal parts of bark and polystyrene, preferably under mist and bottom heat. Cuttings root in about four weeks. Highly recommended for arid fynbos gardens. Plants need pruning after flowering to keep them looking neat and, when grown in summer rainfall areas, they require regular watering in winter and spring. Sow seeds in spring or early summer.

Cherise Viljoen

Freylinia visseri

Common names: honey bell bush, klokkiesbos

Height: up to 3 m

Flowering time: spring to early summer

An erect, vigorous woody shrub with leathery lance-shaped leaves and showy tubular, purplish-maroon flowers borne in clusters. The flowers are attractive to sunbirds and the plant has a persistent rootstock. It thrives in full sun in both sandy or heavy, alkaline and acid soils and is waterwise once fully established and recommended for coastal gardens. Regular pruning after flowering is essential to encourage branching. Propagation is easy from seed or cuttings. Seed should be sown in spring or early summer in a well-drained soil mixture and will germinate within three weeks. Semi-mature cuttings can be taken in autumn, spring or early summer and should be treated with a rooting hormone. Rooting takes place within four weeks. Rooted cuttings should be placed in an area where they can be hardened-off for three weeks before being potted-up.

Cherise Viljoen

Halleria elliptica

Common names: bush honeysuckle, wild fuchsia

Height: 1–2 m

Flowering time: mid-autumn to early spring

A long-lived, small, neat shrub with showy orange, tubular pendulous flowers that attract sunbirds, and ripe berries that are sought after by fruit-eating birds. The elliptic, dark green leaves are toothed along their margins and the plant is ideally suited to shaded or semi-shaded parts of the garden in acid, loamy soil. It can also be successfully displayed as a specimen plant in large containers on shady verandahs. Easily propagated from seed sown in autumn or spring, or from softwood or heel cuttings taken from actively growing shoots in spring, early summer or autumn. They should be treated with a rooting hormone and placed in a propagation unit with intermittent mist and bottom heat of 28°C. Rooting should occur within 6 weeks and newly rooted cuttings require a hardening-off period of 4 weeks.

Halleria lucida

Common names: tree fuchsia, umbinza

Height: up to 12 m

Flowering time: winter to late summer

A long-lived woody shrub or small tree with oval-shaped shiny leaves and tubular, reddish-orange or pale yellow flowers, borne along the stems and trunks. The flowers are visited by sunbirds and the ripe black berries are relished by fruit-eating birds. It is an excellent subject for full sun to densely shaded positions. It requires acid loamy soil, and once established is waterwise. It is easily propagated from seed or cuttings. Seed is sown from spring to mid-summer in a well-drained seedling mix. Germination takes 6 weeks. Softwood or heel cuttings should be taken from actively growing shoots from spring to autumn. They should be treated with a rooting hormone and placed in a propagation unit with intermittent mist and bottom heat of 28°C. Rooting should occur within 6 weeks and newly rooted cuttings require a hardening-off period of 4 weeks.

Selago canescens

Common name: aarbossie

Height: 0.4–0.6 m

Flowering time: late winter to spring

A very showy herbaceous perennial with tufted linear leaves and very small, pale to deep mauve flowers borne in clusters along erect, plume-like inflorescences. It thrives in acid loamy soil and can be grown in both full sun or lightly shaded positions. Once established, the plants are fairly waterwise and it is an excellent subject for mixed fynbos gardens or placed towards the centre of herbaceous borders. The flower-laden branches tend to fall over in windy weather and it is thus most appropriately grown planted closely together in groups, or in mixed plantings with other fynbos species. The plants should be pruned back after flowering. They have a rather short life span and have to be re-propagated every few years by means of seed or cuttings taken in spring. Sow seeds in autumn or spring.

Brenda Szabo

Sutera calciphila

Height: up to 0.1 m

Flowering time: late winter to late spring

A creeping or sprawling soft-leafed perennial with purplish-maroon stems and an abundance of mauve starry flowers with orange perianth tubes. It is an excellent mat-forming groundcover for full sun or lightly shaded positions, ideal for window-boxes and trailing over low retaining walls, or inter-planting with dwarf fynbos bulbs like *Babiana pygmaea* and *Freesia alba*. It thrives in alkaline conditions but easily adapts to heavier loamy, acid soils and is easily propagated by division of rooted trailing stems taken in autumn or spring. Sow seeds in autumn.

Sutera subspicata

Height: 0.2–0.4 m

Flowering time: autumn to late spring

A small compact shrublet with dark green, coarsely toothed leathery leaves and small clusters of pink, mauve or white starry flowers with large yellow anthers, produced in the leaf axils at the tips of branches. It is suitable for windy coastal gardens in acid sandy soil but performs equally well in heavier acid loam, and can be grown in full sun or lightly shaded conditions. It is displayed to best advantage in rock garden pockets, and also makes an attractive container plant. It is propagated by cuttings taken in autumn or spring. Sow seeds in autumn.

Opposite: The flowers of *Freylinia lanceolata* attract insects that provide food for insect-eating birds (see page 133)

Below: *Lobelia valida* in habitat, Cape south coast (see page 142)

Opposite above: *Wahlenbergia undulata* (see page 144)

Opposite below: *Monopsis lutea* (see page 142)

John Winter

11. GROWING FYNBOS LOBELIAS AND CAMPANULAS

The Campanulaceae contains 186 species, including 142 endemics, and is the fourteenth largest family in the Cape Flora.

An analysis of recent germination results for the Campanulaceae showed that seed of the majority of species studied gave a significantly higher germination rate after smoke treatment. It is thus recommended that seed of all species be treated with smoke before sowing.

Ten of the Best Fynbos Lobelias and Campanulas

Liesl van der Walt

Lobelia anceps

Family: Lobeliaceae

Height: up to 0.1 m

Flowering time: early summer to early winter

A delicate, spreading perennial with soft lance-shaped leaves and white, blue or mauve flowers borne singly on long pedicels in the leaf axils. It is an excellent subject for mixed planting with other herbaceous perennials in hanging baskets, large containers or window-boxes. It has a long flowering period and prefers moist, acid, well-drained soil, in sunny or lightly shaded positions. It is best propagated from seeds sown in autumn or spring.

Lobelia comosa

Family: Lobeliaceae

Height: 0.25–0.4 m

Flowering time: spring and summer

A free-flowering perennial occurring in coastal fynbos of the southwestern and southern Cape. It produces erect stems from the base and clusters of bright blue flowers with white markings on the lower tepals. The leaves have a leathery texture and are coarsely toothed along their margins. It requires a composted, well drained acid soil in full sun or light shade and is ideally suited to fynbos borders, rock garden pockets, hanging baskets and mixed containers. The stems should be pruned back after flowering to promote new growth and prevent plants from becoming too woody. Plants are rather short-lived and need to be re-propagated every few years from cuttings taken from strong branch tips in spring and summer, or from seeds sown in autumn or spring.

Lianne Gilmour

Lobelia pinifolia

Family: Lobeliaceae

Common name: wild lobelia

Height: 0.3–0.5 m

Flowering time: throughout the year, but mainly in spring and summer

An attractive, much-branched shrublet producing clusters of bluish-violet, or occasionally pale pink flowers, at the tips of erect branches densely clothed in bright green, linear leaves. It requires acid soil and performs well in full sun positions in containers or rock garden pockets, or placed towards the front of herbaceous borders. It has a long flowering period and the branches should be pruned back lightly in autumn to encourage bushy growth and prevent plants from becoming too woody. It is best propagated from seeds sown in autumn.

Lobelia pubescens

Family: Lobeliaceae

Height: 0.1–0.4 m

Flowering time: late spring to autumn

A low growing, bushy perennial with very short, narrow, greenish-grey leaves and a profusion of white flowers borne singly at the tips of long pedicels. It occurs in coastal habitat in the southern Cape and likes damp but well drained, acid soil in full sun positions. It is ideally suited to containers and rock garden pockets, or used as an edging plant to herbaceous borders. It is best propagated from seeds sown in autumn.

Lobelia valida

Family: Lobeliaceae

Common name: galjoenblom

Height: 0.4 –0.6 m

Flowering time: early summer to mid-autumn

A superb perennial for the garden. The plants branch from the base and the foliage is flat with coarsely toothed margins. Easily grown in sunny positions in light, well-drained soil enriched with compost, this species is ideal for coastal gardens. It prefers a slightly alkaline soil but easily adapts to acid soils and flowers throughout the summer. It is rather short-lived and has to be re-propagated every few years as the plants tend to become woody. They can be propagated by seed or by cuttings. Seed sets freely and can be sown in autumn or spring, germination taking place in a month. Cuttings from strong branch tips are best taken in spring and summer when they are less inclined to rot. Young plants respond well to organic fertilizers and should be encouraged to form bushy growth by pinching out the tips.

Monopsis lutea

Family: Lobeliaceae

Common name: yellow lobelia

Height: 0.2–0.6 m

Flowering time: summer to autumn

This is one of 13 species of *Monopsis* found in central and southern Africa. It is a vigorous, low growing perennial that creeps along the ground, forming dense mats. Its bright yellow, lobelia-like flowers appear in clusters at the tips of the branches and it has a long flowering period throughout summer. It is an excellent groundcover for full sun or semi-shade, preferring damp conditions, and is easily propagated from cuttings taken in autumn or spring, or from division of trailing rooted stems at any time of year. Sow seeds in autumn.

Liesl van der Walt

Monopsis unidentata

Family: Lobeliaceae

Common name: wild violet

Height: 0.2–0.5 m

Flowering time: early to late summer

This fast growing perennial thrives in sunny wet spots and has many long thin stems that root where they touch the ground, forming mats. The small, bright green serrated leaves are grouped near the base of the stems, with the violet-like purple flowers borne at the stem tips. They flower throughout the summer and make an attractive display in hanging baskets, as edging plants or around ponds. The plants die back for a short period in winter and shoot again in spring. They are easily propagated from cuttings taken in autumn or spring, or from division of trailing rooted stems at any time of year. Sow seeds in autumn.

Brian Rycroft

Roella ciliata

Family: Campanulaceae

Height: 0.2–0.5 m

Flowering time: late winter to autumn

A sprawling shrublet from the southwestern Cape with pungent, hairy, needle-like leaves and outstanding large, lilac-blue, cup-shaped blooms (a white form also occurs). The flowers are borne singly at the tips of stiff branches and the tepals have a navy blue band at the base, edged with white on the upper side, and darker blue on the lower side. It is recommended as a rock garden or container subject or as an edging plant, and should be planted in groups to create an effect. It requires acid, well-drained soil in full sun, and once established is very waterwise. The seeds should be sown in autumn but germinate rather slowly and may take up to 3 months. Experiments at Silvermine Nature Reserve showed that seed germination was promoted by plant-derived smoke.

Roella maculata

Family: Campanulaceae

Height: 0.2–0.3 m

Flowering time: midsummer

An erect shrublet branching from the base, with a restricted distribution along the southern Cape coast. It has short, narrow prickly leaves produced in clusters up the stiff hairy stems and bears one to several large blue flowers in a cluster at the tips of the branches. Each tepal has a prominent navy blue marking near the base, edged in white. It requires acid soil and is recommended for sunny rock garden pockets and deep containers, or placed towards the front of a fynbos border. The seeds should be sown in autumn but are slow to germinate and may take up to 3 months. Seed germination is promoted by plant-derived smoke treatment.

Wahlenbergia undulata

Family: Campanulaceae

Common name: giant bell flower

Height: 0.15–0.9 m

Flowering time: late spring to midsummer

A beautiful herbaceous perennial with large mauve, blue or white bell-shaped flowers carried on long wiry pedicels. Each flower lasts just a few days, but a succession of flowers is produced for more than a month. It is best displayed in rock garden pockets or raised beds, and it grows well in full sun or semi-shade in any good garden soil with sharp drainage and regular watering in summer. The plants die back during the winter months before new growth appears again in spring. They are rather short-lived and should be re-propagated every 2–3 years, either by seed or cuttings. Seed is usually faster and is best sown in spring. Good germination is usually obtained from fresh seeds within a few weeks and seedlings should be fed with an organic fertilizer.

Lobelia valida prefers a slightly alkaline soil but easily adapts to acid soils and is ideal for coastal gardens (see page 142)

Below: *Pelargonium cordifolium* can be grown in full sun or lightly shaded positions (see page 158)

Opposite: *Pelargonium betulinum* (see page 157)

12. GROWING FYNBOS PELARGONIUMS

(*Text based on the chapter entitled 'Pelargoniums' in Powrie, 1998*)

The Geraniaceae contains 155 species, including 91 endemics, and is the sixteenth largest family in the Cape Flora.

Our indigenous species of *Pelargonium* have played a vital role in the development of the ornamental hybrid pelargonium (or geranium as it is commonly, but incorrectly known). The introduction of southern African pelargoniums into Europe began in the early seventeenth century with *Pelargonium triste* being one of the first. But it was not until the last decade of the eighteenth century that pelargoniums became popular. William Curtis noted in 1790 that pelargoniums 'show an astonishing readiness to cross and set seed', and this probably marked the beginning of the hybrid pelargonium. Though popularity has fluctuated with time, the variety available and ease of propagation, by vegetative means or from seed, have made the pelargonium a garden favourite, especially for rock garden pockets, window boxes, hanging baskets and for trailing over retaining garden walls.

Pelargonium betulinum is an excellent subject for mass bedding displays (see page 157)

Cultivation

Pelargoniums are found growing in most parts of South Africa in a wide range of habitats, so it is impossible to lay down one set of growing conditions to suit all species. The most important consideration is to note the species' natural habitat, the soil type, the amount of shading, how much rain and when it falls and then to attempt to simulate natural conditions as closely as possible. Many species have wide tolerance limits and are therefore easy to grow.

The basic requirements for growing pelargoniums are good drainage, plenty of light and abundant free air circulation.

Soil

Good sandy loamy soil will suit most pelargoniums. The soil pH is not a critical factor and a neutral to slightly acidic pH is suitable for most species. The pot collection at Kirstenbosch is growing successfully in a mix of 1 part loam, 2 parts sand and 2 parts compost, plus fertilizer. Soil-less media can also be used for pot culture.

Light

Contrary to popular belief, light shade or full shade for part of the day is of benefit to most *Pelargonium* species in South African conditions. In their natural environments they are found growing in the shelter of bushes or rocks where even if the leaves are exposed, the roots are cool. None of the species grow successfully in heavy shade as under these conditions they become very spindly.

Watering

Watering is the most critical factor in controlling the growth conditions of pelargoniums. The amount and frequency of watering depends on the growth medium, climatic conditions, the species requirements and if applicable, the type of container. A general rule of thumb is to only water when dry, as under-watering is preferable to over-watering. For ease of watering it is simpler to group the species according to growth habit.

Geophytic species that have underground storage organs and are seasonally dormant can be divided into winter rainfall species, e.g. *P. rapaceum, P. triste, P. lobatum* and *P. incrassatum* and summer rainfall species, e.g. *P. bowkeri, P. schizopetalum, P. luridum and P. caffrum.* These species require regular water during their growing season but must not be over-watered or the tubers will rot. In their dormant season they require no water and if the storage organ is wet it will rot. Dormant plants can either be stored in their pots, preferably in a cool place or can be lifted, treated with fungicide and stored in a cool dry place in the same way as bulbs. Excessive heat and desiccation will damage the dormant plant, so if it is not possible to store the pots in a cool place they can be cooled by lightly dampening, but not soaking, the surface of the pot.

Succulent and woody species that are dormant or semi-dormant in summer come mainly from the Karoo and other arid areas. They include *P. carnosum, P. crithmifolium, P. praemorsum, P. antidysentericum* and *P. cortusifolium*. These species need a well-drained medium and should not be over-watered, as they are prone to rot. It is advantageous to plant these plants shallowly in their growth medium with a layer of grit on the surface for support as this reduces the chances of rot. Throughout the year, water only when completely dry.

Herbaceous evergreen species that grow all year and come from the south-western and southern Cape include *P. cordifolium, P. fruticosum, P. cucullatum, P. panduriforme* and *P. betulinum*. These species require regular watering to maintain optimal growth but must not be waterlogged. Most of these plants are drought tolerant but will not look their best under drought conditions.

Pruning

Pruning is essential for the vigorous-growing species like *P. denticulatum, P. glutinosum, P. scabrum, P. capitatum* and *P. graveolens*. The plants can be cut back about two thirds to just above a node and any weak growth should be removed .The best time to prune is from late summer to early autumn, which gives the plant time to grow again before winter. If healthy, the prunings can be used for cuttings. To encourage bush formation, it is good practice to pinch out the growing tip of newly potted cuttings. Vigorous species like *P. inquinans, P. salmoneum, P. frutetorum, P. burtoniae* and *P. grandiflorum* can be pinched back regularly during their first year. The succulent and geophytic species do not require pruning except possibly to improve shape or to remove diseased parts. Dead leaves of geophytic species should be cut off, not plucked, as the leaf base protects the growing tip from desiccation and damage.

Feeding

Feeding of potted plants with a balanced fertilizer every tenth watering, when in active growth, is recommended. Some references suggest more regular feeding especially when the plants are growing in a soil-less medium. Care must be take not to over-feed as this can result in lush growth which is prone to disease, or a high salt build-up in the pot medium which has toxic effects. Plants in the garden should be fed at the beginning of their growing season with a balanced slow-release fertilizer.

Potting-on

Potting-on is the moving of a vigorously growing plant to a larger container. This must be done before the plant is root-bound and must be carried out with minimal disturbance to the roots so that growth is not stopped.

Re-potting

Re-potting involves removing the old, exhausted medium from the root ball and re-potting the plant into the same size container. During this process geophytic species may be divided. Roots should be washed clean and any damaged or diseased roots removed. Roots may be treated with a fungicide before re-potting. In the case of the herbaceous evergreen species it is advisable to prune the top portion of the plant as well, to compensate for root loss.

Vegetative Propagation

Stem cuttings

Stem cuttings are the most widely used method of propagating pelargoniums. Plants produced this way will flower within 3–6 months as compared to 12–18 months when grown from seed.

Cuttings can be taken all year round although they root faster in summer and autumn. They should be taken from the terminal growing points of the plants although lower pieces of stem can also be used. The parent plant should be growing vigorously and not be lanky or diseased. Ideally the internodes should be short and the stem fairly firm but not woody. The cutting should contain at least 3–5 leaf nodes and the basal end should be cut neatly just below a leaf node. The leaves and stipules should be carefully removed from the lower two thirds of the cutting leaving a few leaves intact on the top. If the leaves are large, a portion of each leaf may be trimmed off to reduce moisture loss, e.g. *Pelargonium zonale* and *P. inquinans*. Leafless cuttings of the succulent and woody species will root with equal ease. The cuttings should be rooted in trays containing any well-drained medium, e.g. sand covered with a 5 mm thin layer of bark and polystyrene. A rooting hormone powder such as Seradix 2 can be used to improve rooting. The basal tip of the cutting should be dipped in the powder, the cuttings set out in holes in the rooting medium (made using a dibber or a large nail) and the medium should then be pressed firmly around the cutting. The completed tray of cuttings should be watered with a fungicide such as Kaptan. If possible, the cuttings should be placed under mist with bottom heat for 3 or 4 days to give them a chance to settle, then put into a cold frame for the remainder of the rooting period. The cuttings should be watered regularly, and care taken not to over-water.

Pelargonium peltatum is ideal for trailing over retaining walls (see page 161)

Alice Notten

Four to 8 weeks later, after root formation, a feed with Kelpak (or other seaweed extract product) is recommended. One to 2 weeks after this, the cuttings should be ready for potting-up.

Division

Many of the South African pelargoniums are tuberous with either a very short stem or none at all. These species do not lend themselves to the normal vegetative methods of propagation. The tubers that multiply can be divided and planted individually. This method is applied to species such as *P. incrassatum, P. pulchellum* and *P. triste*.

Opposite: *Pelargonium cucullatum* subsp. *cucullatum* (see page 159)

Below: *Geranium incanum* is an excellent groundcover (see page 157)

Unfortunately it is a slow process and is not practical if larger quantities of plants are required quickly.

Propagation from seed

An analysis of recent germination results for the Geraniaceae showed that species could be divided into a number of response groups. Seed of a third of the species of Geraniaceae studied gave significantly higher germination after smoke treatment. Seed of some species studied responded to a hot water treatment. It is thus recommended that both smoke treatment and hot water treatment be combined for seed of all species before seeds are sown.

Pelargonium seed is interesting in that attached to the elliptically shaped seed is a feathered, tail-like structure that is coiled in a spiral arrangement. This tail causes the

seed to twist around in the wind and it drills itself into the soil in a corkscrew fashion, thus ensuring that most of the seeds produced (5 seeds per flower) have a good chance of germinating. For optimum germination, seed is best sown when fresh although it may remain viable for up to 7 years. Sow seed in a light, well-drained soil with a high content of coarse sand. A seed tray 100 mm deep with numerous drainage holes in the base is ideal. Before filling the container, place a layer of roughage or crocks or gravel in the bottom. Firm the soil down gently after levelling it with a piece of wood. The seed should be given a hot water and smoke treatment and then allowed to dry. When dry, broadcast the seed evenly and apply a thin covering of clean white sand. Depth of sowing is usually 1.5 times the size of the seed. Finally, water thoroughly using a fine rose and provide light shade. Germination usually takes place within 10 to 14 days, but may be longer if temperatures are low. The seedlings can be pricked-out into individual containers once they have produced 2 or 3 leaves.

Pelargoniums grown from seed are generally more vigorous and robust than those produced from cuttings. However, plants produced from seed take longer to flower and may also display a certain amount of variation. (The latter characteristic may be desirable if variation is required for selection of superior plant types). For the geophytic species, propagation by seed is the most effective way of producing large numbers of plants. Many of these seedlings will flower within their first year, e.g. *P. oblongatum* and *P. incrassatum*.

A quick guide to germinating pelargonium seed

- Use only mature, plump, fully formed seeds
- Place seeds in hot water at 100°C and then allow to soak for 10 hours
- Then soak in aqueous smoke extract or a commercial smoke seed primer for 24 hours, or smoke seed trays after sowing
- Seeds should be given a light dusting with a fungicide dressing to prevent post-emergence seedling infection
- Sow seeds in a sandy, well-drained soil medium
- Incubate in full sun under autumn temperature conditions, e.g., alternating 10°C (16 hour night) x 20°C (8 hour day)

Opposite above: Ripe seeds of *Pelargonium capitatum* just prior to wind dispersal (see page 158)

Opposite below: *Pelargonium citronellum* has strongly lemon-scented foliage (see page 194)

Using pelargoniums

Pelargoniums are a versatile group of plants that can fill many different roles. As long as soils in which they are grown are well drained, they are quick growing and rewarding plants, offering interesting foliage and showy flowers. The larger species can be used as quick growing shrubs, e.g. *P. papilionaceum, P. citronellum* and *P. cucullatum.* The floriferous smaller species make good bedding plants which, in very harsh areas, can be re-planted annually in much the same way as the cultivars are treated in Europe, e.g. *P. inquinans* and *P. peltatum.* A herb garden is enhanced by the addition of scented pelargoniums like *P. tomentosum, P. crispum* and *P. graveolens.* A well-drained rockery would provide a suitable home for the more succulent species, e.g. *P. magenteum* and *P. fulgidum.* The geophytic species may be used in the same way as bulbs either in mass displays or in rockery pockets, e.g. *P. oblongatum, P. incrassatum* and *P. auritum.*

All the smaller, succulent, scented and geophytic species make good pot subjects for the, patio or window-sill. Make sure that the species chosen is placed in a position that it can tolerate - whether it be full sun or partial shade.

Pests and diseases

While pelargoniums are regarded as tough plants, they are not without their problems. Many grow happily in the open, free of disease, but when grown indoors or in a greenhouse, the warmer conditions and lack of free air circulation encourage pests, and the application of pesticides becomes necessary. The major pests, their symptoms and control are as follows:

Aphids are small green insects, winged or wingless, that attack stem tips sucking plant sap and causing wilt. Treat with Pirimor, Malathion or Ripcord.

Caterpillars of various species chew leaf margins, and are evident by their droppings. Control with Malathion or Karbaspray.

Red spider mite is visible as tiny red dots on leaves. Leaves become mottled, yellow and shrivelled. Treat with Garden Gun or Oleum.

Snails chew shallow holes in fleshy pelargoniums and are evident by their slime trail. There are many snail baits on the market.

Whitefly are small white, winged insects found on the underside of leaves, sucking sap and causing leaf curl. Treat with Ripcord, Garden Gun and yellow sticky traps.

Rust is a fungus that appears as brown spots on the underside of leaves and yellow spotting on the upper surface. Treat with Cupravit, Koprox, Biltox, Funginex or Dithane.

Powdery mildew, also a fungus, is a whitish, powdery growth on leaves that can be controlled by Benlate, Funginex, copper-oxychloride or a dusting with Flowers of Sulphur.

Oedema, a swelling and splitting of stems that creates an entry site for disease, which is caused by over-watering, can be prevented by reducing the frequency of watering.

Ten of the Best Fynbos Geraniums and Pelargoniums

(see also Appendix 2, page 194)

Geranium incanum

Common names: horlosies, vrouebossie

Height: 0.2–1 m

Flowering time: late winter to early summer

A beautiful spreading perennial with a profusion of large mauve, pink or white flowers and attractive, finely divided leaves. This long-lived plant has a thickened taproot and is an excellent mat-forming groundcover for sprawling over rock garden pockets and low retaining walls, or as a subject for window-boxes and hanging baskets. It requires full sun to flower well and thrives in acid loam. It is an ideal companion plant for smaller restios like *Elegia persistens* and *Elegia stipularis*. It is fast growing and, once established, is very waterwise. Easily propagated by division of thick clumps in autumn or early spring, or from seeds sown in autumn or spring.

Liesl van der Walt

Pelargonium betulinum

Common name: maagpynbossie

Height: up to 0.5 m

Flowering time: spring to summer

One of the most desirable and easily grown pelargoniums, this shrublet becomes covered with clusters of pinkish-mauve or white flowers and has attractive oval-shaped, leathery leaves. It forms a rounded bush and is an excellent subject for mass bedding displays, rock garden pockets or mixed fynbos beds. It is especially showy when inter-planted with restios like *Chondropetalum tectorum* and *Thamnochortus insignis*. It requires full sun and thrives in almost any well-drained soil. The branches should be pruned back after flowering to encourage compact bushy growth, and it is easily propagated from cuttings taken in autumn or early summer, or from seeds sown in autumn or spring.

Pelargonium capitatum

Common names: rose-scented pelargonium, kusmalva

Height: 0.2–0.5 m

Flowering time: early to late spring

An easily grown, rambling shrublet with fragrant, heart-shaped leaves and compact heads of pinkish-mauve flowers produced at the tips of long hairy peduncles. It is an excellent subject for difficult windy, sandy coastal gardens and thrives in alkaline and acid soils, in full sun positions. Suited to rock garden pockets, pavement beds or mixed fynbos plantings, and very waterwise. Easily propagated from cuttings taken in autumn or early summer, or from seeds sown in autumn or spring.

Pelargonium cordifolium

Common name: heart-leafed pelargonium

Height: up to 1 m

Flowering time: spring to early summer

A small shrub with attractive, aromatic, deep green, heart-shaped leaves and showy large pink flowers. It is an ideal plant for mixed borders and rock gardens that receive water all year round, and suitable for full sun or lightly shaded positions. This species requires well composted, well drained soil. Branches should be pruned back after flowering to encourage bushy growth. The form known as *P. cordifolium* 'Valentine' has eye-catching, deep red stems and large pink flowers. Easily propagated from cuttings taken in autumn or early summer, or from seeds sown in autumn or spring.

Liesl van der Walt

Pelargonium cucullatum subsp. *cucullatum*

Common name: wildemalva

Height: up to 2 m

Flowering time: spring to summer

A vigorous, fast-growing shrub with clusters of large mauve or purple flowers and rounded, leathery leaves with jagged, edges. It is ideally placed towards the rear of herbaceous borders or planted in mass displays or in mixed fynbos beds. Its sturdy branches should be pruned back regularly to prevent plants from becoming too woody. It tolerates a wide range of growing conditions, from sandy alkaline soils to acid loam, and is waterwise once established. *P. cucullatum* subsp. *tabulare* has large, rounded, scented leaves. Both are easily propagated from cuttings taken in autumn or early summer, or from seeds sown in autumn or spring.

Pelargonium fulgidum

Common name: rooimalva

Height: 0.2–0.4 m

Flowering time: winter to summer

A rambling shrublet with silky, hairy, greenish-grey leaves and small bright red flowers borne on long spreading pedicels. The branches have distinct swollen nodes and the leaves are deciduous, the old ones falling off in summer and new leaves developing in early autumn. It requires full sun and prefers acid, well-drained, humus-rich soil and is an ideal rock garden or container plant. Care must be taken to reduce watering considerably in summer, failing which the rootstock may rot. Easily propagated from cuttings taken in autumn, or from seeds sown in autumn.

Pelargonium incrassatum

Height: 0.2–0.3 m

Flowering time: early to late spring

A beautiful geophytic, winter-growing species with a rosette of feathery grey-green leaves and rounded heads of bright magenta flowers borne on long, slender peduncles. The tuberous rootstock is planted with the apex at or just below soil level and it is most suitably grown as a container plant in a well drained, gritty medium, in a sunny position. Watering is required during the winter growing period but the soil medium must be kept completely dry from early summer to early autumn. Easily propagated from seeds sown in autumn.

Pelargonium inquinans

Common names: scarlet geranium, wildemalva

Height: 1–2 m

Flowering time: throughout the year, with a peak in spring

A long-lived, small shrub from the western and eastern parts of the Eastern Cape with bright red flowers and rounded, softly hairy leaves. It is an excellent choice for mixed herbaceous plantings and can be grown in full sun or lightly shaded positions. The branches should be pruned back regularly to encourage bushy growth and prevent plants from becoming too woody. Easily propagated from cuttings taken in autumn or early summer, or from seeds sown in autumn or spring.

Pelargonium peltatum

Common name: ivy-leafed pelargonium

Height: 0.1–2 m

Flowering time: throughout the year, with a peak from spring to midsummer

A superb herbaceous perennial climber with ivy-like leaves and pale pink, mauve or purple blooms with dark magenta markings. It is ideal for hanging baskets, window-boxes and mixed planters, or for trailing over retaining walls or even as a groundcover. The vigorous form known as *Pelargonium peltatum* 'Worcester' has much smaller, prominently marked leaves and showy mauve flowers (see page 240). This species does well in coastal areas and thrives in alkaline and acid soils. Easily propagated from cuttings taken in autumn or early summer, or from seeds sown in autumn or spring.

Pelargonium quercifolium

Common name: muishondbos

Height: 1–1.5 m

Flowering time: early spring to midsummer

A robust small shrub with fragrant, hairy foliage and small bright pink flower heads marked with dark magenta. The plants are fairly waterwise once established and are ideal subjects for large rock gardens, mass plantings or mixed fynbos beds. They must have well-drained soil and tolerate acid and alkaline conditions. Should be pruned back heavily after flowering to encourage bushy new growth. Easily propagated from cuttings taken in autumn or early summer, or from seeds sown in autumn or spring.

Below: *Podalyria calyptrata* has sweet-scented flowers and is useful for planting along boundaries as an informal hedge (see page 167)

Opposite: *Liparia splendens* (see page 167)

13. GROWING FYNBOS LEGUMES

The Fabaceae is the second largest family in the Cape Flora and contains 760 species, including 627 endemics.

An analysis of recent germination results for the Fabaceae showed that a quarter of the species tested responded to smoke treatment, but the most effective dormancy-breaking treatment was a treatment using hot water at 100°C that softens or fractures the hard seed coats.

A quick guide to germinating legume seed

- Use only mature, plump, fully formed seeds
- Seeds have hard, impermeable seed coats that require softening or fracturing, so place them in hot water at 100°C and allow them to soak for 10 hours
- Then soak them in aqueous smoke extract or a commercial smoke seed primer for 24 hours or smoke seed trays after sowing
- Give seeds a light dusting with a fungicide dressing after the soaking treatment to prevent post-emergence fungal infection of seedlings
- Sow seeds in a sandy, well-drained soil medium
- Incubate in full sun under autumn temperature conditions, i.e. alternating 10°C (16 hour night) x 20°C (8 hour day)

This group of plants is fast-growing, generally not long-lived, surviving between 5–15 years depending on the species.

Left: The beautiful silvery foliage and small pink flowers of *Podalyria sericea* (see page 168)

Below: *Podalyria sericea* requires a sunny position in well-drained, acid, loamy soil (see page 168)

Ten of the Best Fynbos Legumes

(see also Appendix 2, page 194)

Liesl van der Walt

Cyclopia genistoides

Common names: honeybush tea, heuningtee

Height: 1.5–2 m

Flowering time: spring to early summer

A fast-growing, densely-branched, erect small shrub from the southwestern and southern Cape, with clusters of needle-like leaves and showy terminal racemes of bright yellow 'pea' flowers borne on short pedicels. It requires deep acid soil, plenty of moisture in winter and a full sun position to flower well. It is highly recommended for mixed fynbos beds, and the branches should be pruned back heavily after flowering to encourage bushy new growth. Best propagated from seeds sown in early spring.

Cyclopia sessiliflora

Common name: Heidelbergtee

Height: 1–1.3 m

Flowering time: autumn to spring

A magnificent erect shrub with long, willowy branches bearing clusters of needle-like leaves and masses of striking bright yellow 'pea' flowers produced in the leaf axils. The plants require well-drained, deep acid soil in full sun and are excellent subjects for mixed fynbos gardens. The branches should be pruned back heavily after flowering to encourage bushy new growth and prevent plants from becoming too woody. Best propagated from seeds sown in early spring.

Indigofera brachystachya

Common names: beesbossie, nentabossie

Height: 0.5–1.2 m

Flowering time: early summer to spring, with a peak in late winter

A dense small shrub with short, hairy greenish-grey leaves and masses of small pink or mauve flowers borne on short peduncles in the leaf axils. It thrives in alkaline and acid, well-drained soils and is an excellent subject for difficult, windy coastal gardens in full sun positions, and is recommended for rock garden pockets and mixed fynbos beds. Best propagated from seeds harvested from the attractive hairy pods, sown in autumn or spring.

Liesl van der Walt

Indigofera frutescens

Common names: river indigo, kouebos

Height: 2–7 m

Flowering time: late winter to early autumn

A graceful, much-branched shrub or small tree with dark green or grey-green compound leaves and showy sprays of pale pink, mauve or red flowers produced on short peduncles in the leaf axils. It produces attractive small brown seed pods and is an excellent subject for acid soils in small and large gardens. Light pruning of the branches after flowering encourages bushy growth and it is easily raised from seeds sown in autumn or spring.

Liesl van der Walt

Liparia splendens

Common name: mountain dahlia

Height: 0.6–1 m

Flowering time: autumn to midsummer

A striking, long-lived small shrub for the fynbos connoisseur. It has rounded yellowish-orange, pendulous flower heads and is endemic to mountain and lowland fynbos in the south-western Cape. It requires full sun and an acid soil, and is best displayed in sloping rock garden pockets. The branches should be pruned back regularly after flowering to encourage bushy new growth, and it readily re-sprouts from the base. Best propagated from cuttings taken in autumn or spring, or from seeds sown in autumn or spring.

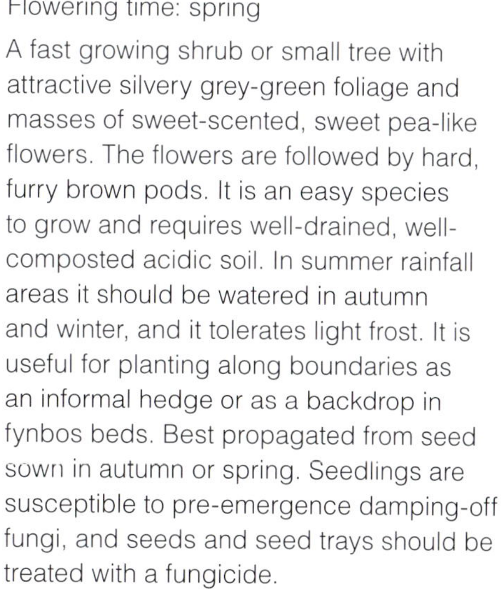

Podalyria calyptrata

Common names: sweet pea bush, keurtjie, ertjiebos

Height: shrub: 2–3 m; small tree: 4–5 m

Flowering time: spring

A fast growing shrub or small tree with attractive silvery grey-green foliage and masses of sweet-scented, sweet pea-like flowers. The flowers are followed by hard, furry brown pods. It is an easy species to grow and requires well-drained, well-composted acidic soil. In summer rainfall areas it should be watered in autumn and winter, and it tolerates light frost. It is useful for planting along boundaries as an informal hedge or as a backdrop in fynbos beds. Best propagated from seed sown in autumn or spring. Seedlings are susceptible to pre-emergence damping-off fungi, and seeds and seed trays should be treated with a fungicide.

Podalyria sericea

Common name: silky podalyria

Height: 0.5–1 m

Flowering time: early to midwinter

An attractive, rounded, small shrub with silky, silvery foliage and small, pink flowers produced singly in the leaf axils. It requires a sunny, well-drained position in acid loamy soil and is best displayed in rock garden pockets or towards the front of mixed fynbos beds. Its silvery foliage provides excellent contrast against green-leafed fynbos plants and it is fairly waterwise once fully established. Easily raised from seeds sown in autumn or spring.

Psoralea pinnata

Common names: fonteinbos, penwortel

Height: 2.5–4 m

Flowering time: early spring to mid-autumn

An erect, re-seeding or re-sprouting, willowy shrub or small tree with spiky linear leaves and clusters of bright blue 'pea' flowers with white keel petals. It occurs in marshy areas and on stream banks in mountain and lowland fynbos in the southwestern Cape, and is highly recommended for mixed fynbos gardens with very acid soils. It likes full sun and the branches should be pruned back after flowering to prevent plants from becoming too woody. Easily raised from seeds sown in autumn or spring.

Sutherlandia tomentosa

Common names: eendjies, kippiebos, rooikeurtjie

Height: 1–1.5 m

Flowering time: early to late spring

A small rounded shrub with very attractive silvery-grey leaves and clusters of pendulous, bright red 'pea' flowers. It produces large, rounded, pinkish-red, inflated seed pods and is an excellent waterwise plant for sandy coastal gardens in full sun. It is fast growing and can be grown in acid or alkaline soils. It is displayed to best advantage in rock garden pockets or placed towards the front of fynbos borders. Pruning of the branches after flowering encourages bushy, compact new growth and extends the life of the plant. Highly prized by nectar-feeding birds, especially sunbirds. It is best propagated from seeds sown in autumn or spring.

Virgilia divaricata

Common names: pink blossom tree, keurboom

Height: 15–20 m

Flowering time: late winter to early summer

A fast-growing, small tree, covered in dense sprays of sweetly scented, deep pink blooms. It thrives in both sandy and heavy soils and once established, is waterwise. Recommended for small and large gardens, it is an excellent species for quickly establishing a screen against unsightly structures, or used as temporary protection for smaller, slow-growing trees. The flowers attract carpenter bees that are their pollinators, and it is a highly favoured nesting tree for garden birds. Easily raised from seed which has been soaked in boiling water and sown in spring.

Below: Fruiting branches of *Berzelia abrotanoides.* It requires full sun and a moist, well-drained, acid soil (see page 173)

Opposite: *Brunia laevis* (see page 175)

14. GROWING FYNBOS BRUNIAS

(This chapter incorporates information from articles originally written by Fiona Powrie and Roger Tuckledoe)

The Bruniaceae is one of the distinctive families of the Cape Flora and is mostly endemic. It has an estimated 64 species in 11 genera and only 3 species extend outside the confines of the Cape Region. About a third of the family are rare in the wild because they have very localized distributions, or are being over exploited for the cut-flower trade or because of the encroachment of alien vegetation.

Although the Bruniaceae is not entirely endemic to the Cape Region, the other 6 rare and endemic fynbos families, the Grubbiaceae, Retziaceae, Roridulaceae, Pennaeaceae, Stilbaceae and Geissolomaceae are, and they are often found with the Bruniaceae in the wild.

Brunias, berzelias and nebelias are what might be considered typical of the Bruniaceae. They are easily rccognized by their masses of fluffy, rounded flower heads that are usually white or cream, as in *Brunia noduliflora* (see page 176), although other colours also occur, like the bright red of *Brunia stokoei* (see page 176).

However, many of the Bruniaceae look like something else – the staavias look like daisies and *Audouinia capitata* looks like an erica, hence its common name 'false heath' (see page 3). Linconias and thamneas also look like ericas, while lonchostomas could as easily be mistaken for *Struthiola* (family Thymelaeaceae). *Raspalia*, *Pseudobaeckea* and *Mniothamnea* look like very scruffy phylicas (family Rhamnaceae).

The most distinctive characteristics of the Bruniaceae are the black dot at the tip of their leaves and their characteristic coffee smell, both clearly discernible in *Brunia albiflora* (coffee bush) (see page 174).

Members of the Bruniaceae tend to grow on the cooler southern mountain slopes or in seeps, and therefore benefit from regular watering throughout the year.

Some members of the family form recognizable patterns with others. For example, on the southern end of all populations of *Staavia dodii* (diamond eyes) one finds *Audouinia capitata*.

An analysis of recent germination results for the Bruniaceae showed that nearly half the number of species tested responded favourably to smoke.

Below: *Berzelia lanuginosa* under cultivation at Kirstenbosch (see page 174)

Ten of the Best Fynbos Brunias

Berzelia abrotanoides

Common name: redleg buttonbush

Height: 1–1.5 m

Flowering time: late winter to early summer

An upright leafy shrub with dense clusters of bright green immature flower heads (above) that transform to cream when open and to red when mature (see page 170), produced on prominent red peduncles at the branch tips. The plants re-sprout from a woody base after wild fires and it requires moist but well-drained, acid soil in full sun positions. It is recommended for mixed fynbos beds and the branches should be pruned back lightly after flowering to encourage bushy growth. Cut stems make interesting, long-lasting additions to fynbos arrangements. Seeds should be smoke-treated before sowing in autumn.

Berzelia galpinii

Height: 1.5–2 m

Flowering time: spring to early summer

An upright, many-branched shrub with dense clusters of beautiful large, spherical, cream-coloured flower heads with very long stamens, produced at the tips of branches. It requires deep acid, sharply drained soil in full sun and plenty of water during winter. The flowers attract large, metallic-green scarab beetles and it is recommended for mixed fynbos gardens. Branches should be pruned back lightly after flowering to encourage bushy new growth. It makes a good cut-flower. Smoke-treat seeds before sowing in autumn.

Berzelia lanuginosa

Common names: common buttonbush, vleiknoppiesbos

Height: 1.5–2 m

Flowering time: winter to early summer

A densely branched, small shrub with soft, needle-like leaves and cream flowers in showy, globular flower heads. It requires a moist but well-drained, lime-free medium that is rich in organic matter, in full sun or partial shade. It makes a good cut-flower and is an excellent subject for mixed fynbos beds. Propagation is most successful by seed sown in autumn, or alternatively, from cuttings taken from late summer to autumn. Seed should be dusted with a fungicide to prevent pre-and post-emergence fungal attack, and responds well to smoke treatment, germination taking 6 to 8 weeks. Semi-hardwood cuttings 8–10 cm long should be placed in a rooting medium of 50:50 milled bark and polystyrene, treated with a rooting hormone and placed on heated benches with an overhead mist spray. Success with cuttings is usually not as great as with seed. Branches should be pruned back lightly after flowering to maintain a neat shape.

Brunia albiflora

Common names: coffee bush, stompies

Height: 2–3 m

Flowering time: late summer to autumn

A striking, densely leafy, single-stemmed, tall shrub with flat-topped clusters of spherical white flower heads that are coffee scented. The ripe fruits add to its appeal and it is a fast-growing species that likes deep, acid soil in full sun and moist conditions throughout the year. It is highly recommended for mixed fynbos gardens and is one of the most rewarding brunias to grow. The branches should be pruned back lightly after flowering to encourage bushy new growth. It is eaily propogated from semi-hardwood cuttings 8–10 cm ling, taken from late summer to mid-winter. Smoke-treat seeds before sowing in autumn.

Brunia alopecuroides

Height: up to 1–1.5 m

Flowering time: spring to early summer

A densely leafy, sturdy, much-branched shrub with narrow, lance-shaped leaves and showy, small heads of cream-coloured flowers. The long-lasting fruits are equally attractive, first turning deep red and finally to brown as they mature. It is an easily cultivated species requiring a full sun position in acid soil, and needs regular deep watering throughout the year. Highly recommended for mixed fynbos beds, and a good cut-flower. Smoke-treat seeds before sowing in autumn.

Brunia laevis

Common name: grey snowbush

Height: 0.9–1.5 m

Flowering time: mid to late summer

A sturdy, rounded grey-green fynbos shrub with small overlapping, oblong shaped leaves closely adhering to the branches. The flower heads are crowded into dense rounded clusters at the tips of branches and produce cream-coloured individual flowers with prominent long stamens. The eye-catching, grey, unopened flower heads are just as attractive and highly sought after for the vase. The plants coppice from a persistent woody base and require well drained, acid soil in full sun and regular watering throughout the year. Smoke-treat seeds before sowing in autumn.

Brunia noduliflora

Common names: common snowbush, fonteinbossie, stompie

Height: 0.8–1 m

Flowering time: autumn to midwinter

An upright, rounded fynbos shrub with erect branches clothed in soft hairs and very short, closely pressed, lance-shaped leaves. The cream-coloured flower heads are crowded into dense rounded clusters at the tips of branches and are especially showy due to their very long stamens. The plants re-sprout from a persistent woody base and are excellent subjects for mixed fynbos gardens in acid, well drained soil, in full sun. They require regular watering throughout the year. It is a good cut-flower. Smoke-treat seeds before sowing in autumn.

Peter Jackson

Brunia stokoei

Common name: rooistompie

Height: 1–2 m

Flowering time: late summer to autumn

One of the most striking members of the Bruniaceae, this slender rigid shrub bears deep red flowers in large heads produced in clusters at the tips of major branches. The erect sturdy branches are clothed in relatively broad, smooth, needle-like leaves and the plant re-sprouts from a persistent subterranean rootstock after fire or heavy pruning. It requires well-drained, acid soil in sunny positions and regular watering throughout the year, and is best suited to mixed fynbos beds. Best propagated from smoke-treated seeds sown in autumn.

Nebelia paleacea

Height: 1–1.5 m

Flowering time: late spring to autumn

A very floriferous and attractive rounded shrub with erect branches densely covered in very short needle-like leaves. The small cream-coloured flowers are produced in groups of dense heads at the tops of the branches and the plants re-sprout from a subterranean rootstock (lignotuber) after veld fires or heavy pruning of the branches. It requires deep, well-drained acid soil in full sun positions and regular watering throuoghout the year. It is best suited to mixed fynbos beds and rock gardens. Smoke-treat seeds before sowing in autumn.

Fiona Powrie

Staavia glutinosa

Common names: fly-catcher bush, vlieëbos

Height: 1–1.5 m

Flowering time: late winter to spring

A striking small shrub with a single basal stem, erect slender branches clothed in needle-like leaves, and blackish flower heads surrounded by eye-catching white bracts. It requires acid, well drained soil in a sunny position and the branches should be pruned back heavily after flowering to encourage bushy new growth. It has been successfully propagated at Kirstenbosch by cuttings taken from the previous season's growth in early summer, just as new vegetative growth begins. Cuttings can also be taken from vigorous re-sprout growth that develops from the persistent rootstock after pruning. It can also be grown from smoke-treated seeds sown in autumn, and is a good cut-flower.

Below: Regeneration and profuse flowering in *Erica cerinthoides* following a wild fire (see page 68)

Opposite: Burnt cones of *Protea repens* (see page 21)

15. SMOKE TREATMENT FOR SEEDS

Very promising results have been obtained, showing significantly improved seed germination in many species following treatment with smoke. Approximately 283 species from the Proteaceae, Ericaceae, Restionaceae, Bruniaceae, Asteraceae, Fabaceae, Aizoaceae, Poaceae (grasses), Rutaceae (buchus), Geraniaceae (pelargoniums) and other families have been screened for a response to smoke.

In the plant nursery, the procedure for smoking seed is a relatively simple one and is easy to carry out.

Seed is sown in conventional plastic trays and is covered by a thin layer of soil or finely milled bark. The trays are placed in a polythene tent and smoke is pumped into the tent by means of a plastic pipe from a large metal drum (see page 181). The smoke is generated in the drum by burning a mixture of dry and green fynbos leaf and stem material. The trays are left in the smoke for 1–2 hours. At the end of this period the trays are removed and the seeds carefully watered to wash the smoke deposit into the soil.

The seed trays are then placed under cover in a shade house until the seeds have germinated. Seeds of many fynbos species require fluctuating day/night temperatures for germination (e.g. 20°C day/10°C night). This is a germination cue related to the post-fire environment in the natural habitat, which has a Mediterranean-type climate, with summer drought and winter rain. Removal of the vegetation cover by fire in the late summer or early autumn results in more extreme day and night soil temperatures at a time when the first rains are likely to begin. The best time to sow and smoke-treat fynbos seeds is thus in the late summer and early autumn (March to May in the Southern Hemisphere).

Smoke in a packet

In addition to plant-derived smoke itself, aqueous extracts of smoke also break seed dormancy and give the same dramatic improvement in seed germination in many species which have previously been difficult or impossible to germinate in the nursery. In order to make smoke technology available to a wider spectrum of botanists, horticulturists and gardening enthusiasts, Kirstenbosch researchers developed a seed primer incorporating aqueous smoke extracts. Absorbent paper discs are impregnated with aqueous smoke solution. The discs are then dried and packed into polythene packets ('Smoke in a packet'). The smoke seed primer can be used by gardeners to break dormancy in seeds and thus germination can be improved without having to light a fire. In order to activate the primer, water is added to the paper and seeds are soaked in the solution for 24 hours. Several other smoke seed primers are now also available commercially.

Typical of the response from users of smoke seed primers was a report by Pienaar (1995) who had previously achieved considerable success with the germination of seed of most vygie species (Aizoaceae) from the drier regions of South Africa. He stated that the seed primer should be used by all horticulturists interested in growing vygies from seed collected from such locations. He concluded that, 'This discovery has definite survival value for those rare, localized and endangered fynbos taxa and cannot be doubted. It is a major breakthrough in the *ex situ* cultivation of such species and will ensure their continued survival'.

Above: Smoke discs are available commercially in several different colours depending on supplier

Opposite above: Smoke treatment of newly sown fynbos seeds

Opposite below: Smoke promotes germination in seeds of *Leucadendron tinctum* and germination is further improved if smoke is combined with a light scarification (see page 33)

References and Further Reading

General

Coetzee, C. (Ed) (1996). *Protea Cultivation.* Agricultural Research Council, Elsenburg (unpublished manual).

Cowling, R. and Richardson, D. (1995). *Fynbos, South Africa's unique floral kingdom.* Fernwood Press, Cape Town.

Du Plessis, N.M. and Duncan, G.D. with watercolours by Elise Bodley (1989). *Bulbous Plants of Southern Africa: a guide to their cultivation and propagation.* Tafelberg Publishers, Cape Town.

Germishuizen, G. and Meyer, N.L. (Eds) (2003). *Plants of Southern Africa: an annotated checklist. Strelitzia* 14. National Botanical Institute, Pretoria.

Goldblatt, P. and Manning, J.C. (2000). *Cape Plants, A Conspectus of the Cape Flora of South Africa. Strelitzia 9.* National Botanical Institute, Pretoria, South Africa and MBG Press, Missouri Botanical Gardens, St Louis, Missouri, USA.

Haaksma, E.D. and Linder, H.P. (2000). *Restios of the Fynbos.* Botanical Society of South Africa, Cape Town.

Jeppe, B.J. and Duncan, G.D. (1989). *Spring and Winter Flowering Bulbs of the Cape.* Oxford University Press, Cape Town.

Knight, J. (1809). *On the cultivation of the plants belonging to the Natural Order of Proteeae.* William Savage, London. http://protea.worldonline.co.za/growknight.htm

Linder, H.P. (2002). The African Restionaceae. An intkey identification and description system. *Contributions from the Bolus Herbarium, University of Cape Town*, vol 20.

Matthews, L. J. (1993). *Proteas of the World.* Bok Books, Durban.

Powrie, F. (Ed) (1998). *Grow South African Plants.* Kirstenbosch Gardening Series. National Botanical Institute, Cape Town.

Rebelo, A. (1995). *Sasol Proteas: A field guide to the proteas of southern Africa*, Fernwood Press in association with the National Botanical Institute, Cape Town.

Van Staden, J., and Brown, N.A.C. (1977). Studies on the germination of South African Proteaceae: A review. *Seed Science and Technology* 5: 633-643.

Vogts, M. (1989). *South African Proteaceae: Know them and grow them.* Struik, Cape Town.

Bulbs

Duncan, G.D. (1982). Ten *Babiana* species for promotion. *Veld & Flora* 68(2): 47-48.

Duncan, G.D. (1988). *The Lachenalia Handbook. Annals of Kirstenbosch Botanical Gardens,* vol. 17. National Botanical Gardens, Cape Town.

Duncan, G.D. (1996). *Growing South African bulbous plants.* National Botanical Institute, Cape Town.

Duncan, G.D. (1999). Ixias for pot and garden. *Veld & Flora* 85(2): 78-79.

Duncan, G.D. (2000). *Grow Bulbs.* Kirstenbosch Gardening Series. National Botanical Institute, Cape Town.

Duncan, G.D. (2000). Wine cups. *Veld & Flora* 86(3): 114-117.

Duncan, G.D. (2001). *Chasmanthe. Veld & Flora* 87(3): 108-111.

Duncan, G.D. (2002). *Grow Nerines.* Kirstenbosch Gardening Series. National Botanical Institute, Cape Town.

Duncan, G.D. (2002). Dwarf watsonias: ten of the best for pots and rock gardens. *Veld & Flora* 88(3): 94-98.

Duncan, G.D. (2003). Endangered geophytes of the Cape Floral Kingdom. *Curtis's Botanical Magazine* 20(4): 245-250.

Duncan, G.D. (2004). Amaryllis Magic. *Veld & Flora* 90(4): 143-147.

Duncan, G.D. (2004). The genus *Veltheimia. The Plantsman New Series* 3(2): 97-103.

Duncan, G.D. (2004). The new Kirstenbosch bulb terrace. *Veld & Flora* 90(2): 62-65.

Duncan, G.D. (2005). *Haemanthus and their cultivation. The Plantsman New Series* 4(4): 220-226.

Duncan, G.D. and Joubert, E. (2005). *Ixia viridiflora. Flowering Plants of Africa* 59: 36-42.

Pelargoniums

Bayer, M.B. (1989). *Pelargonium.* In: Du Plessis, N.M. and Duncan, G.D., *Bulbous Plants of Southern Africa*: 53-57. Tafelberg, Cape Town.

Miller, D. (1996). *Pelargoniums: a gardeners guide to the species and their hybrids and cultivars.* B.T. Batsford, London.

Van der Walt, J.J.A. (1977). *Pelargoniums of southern Africa,* vol. 1. Purnell, Cape Town.

Van der Walt, J.J.A. and Vorster, P.J. (1981). *Pelargoniums of southern Africa,* vol. 2. Juta, Cape Town.

Van der Walt, J.J.A and Vorster, P.J. (1988). *Pelargoniums of southern Africa, including some representatives from other parts of the world,* vol. 3. National Botanic Gardens, Cape Town.

Webb, W.J. (1984). *The pelargonium family: the species of* Pelargonium, Monsonia *and* Sarcocaulon. Croom Helm, London.

Smoke and seed germination

Brown, N.A.C., van Staden, J., Daws, M.I. and Johnson, T. (2003). Patterns in the seed germination response to smoke in plants from the Cape Floral Region. *South African Journal of Botany* 69: 514-525.

Brown, N.A.C. and Botha, P.A. (2004). Smoke Seed Germination Studies and A Guide to Seed Propagation of Plants from the Major Families Of The Cape Floristic Region, South Africa. *South African Journal of Botany* 70(4): 559-580.

Brown, N.A.C. (1993). Promotion of germination of fynbos seeds by plant-derived smoke. *The New Phytologist* 123: 575-583.

Brown, N.A.C. and Botha, P.A. (2002). Smoking seeds. *Veld & Flora* 88: 68-69.

Brown, N.A.C., Botha, P.A. and Prosch, D.S. (1995). Where there's smoke. *The Garden: Journal of the Royal Horticultural Society* 120: 402-405.

Brown, N.A.C., Jamieson, H. and Botha, P.A. (1994). Stimulation of seed germination in South African species of Restionaceae by plant-derived smoke. *Plant Growth Regulation* 15: 93-100.

Brown, N.A.C., Kotze, G., and Botha, P.A. (1993). The promotion of seed germination of Cape *Erica* species by plant-derived smoke. *Seed Science and Technology* 21: 179-185.

Brown N.A.C. and Van Staden, J. (1997). Smoke as a germination cue: A review. *Plant Growth Regulation* 22:115-124.

Brown, N.A.C., Van Staden, J. and Brits, G.J. (1996). Propagation of Cape Proteaceae, Ericaceae and Restionaceae from seed. *Combined Proceedings of the International Plant Propagator's Society* 46: 23-27.

De Lange J.H. and Boucher C. (1990). Autecological studies on *Audouinia capitata* (Bruniaceae). I. Plant-derived smoke as a seed germination cue. *South African Journal of Botany* 56: 700-703.

Pienaar, U. de V. (1995). Instant smoke seed primer stimulates the germination of vygie seeds. *Veld & Flora* 81: 93.

Van Staden, J., Brown, N.A.C., Jager, A.K. and Johnson, T.A. (2000). Smoke as a germination cue. *Plant Species Biology* 15: 167-178.

Watsonia borbonica subsp. *ardernei* is ideally suited to interplanting with spring-flowering fynbos annuals (see pages 86, 87 and 115)

GLOSSARY

butenolide a compound in smoke from burning plant material that promotes seed germination

corymb a type of inflorescence, a raceme in which lower flower stalks are proportionally longer, forming a flattish top

dehiscent seed released from the ovary after maturation

dioecious with unisexual flowers, the male and female flowers on separate plants

diurnal of each day

elaiosome the fleshy structure on ant-dispersed protea fruits which is consumed by the ants. It may be a fleshy lump at the base of a fruit or it may cover the entire fruit

floret small flower, especially the flower of daisies; disc florets: the central florets of a daisy, having 5 small equal petals each; ray florets: the outer florets of a daisy, having a single very large strap-shaped petal-like lobe each

fynbos a derivative of the Dutch 'fijnbosch', referring to the community of small shrubs, evergreen and herbaceous plants and bulbs confined to the south-western and southern Cape

indehiscent fruit remaining closed at maturity, e.g. one-seeded fruit or nut

lignified woody

myrmecochory seed dispersal by ants, in which ants carry fruits underground into their nest to eat the elaiosome surrounding the fruit

offset a short shoot that arises from an axillary bud near the base of a stem and gives rise to a daughter plant

pappus a modified calyx made up of a ring of fine hairs or scales that aid the wind dispersal of the fruit – sometimes forming a parachute-like structure

pedicel the stalk of an individual flower

peduncle the stalk of a flower or fruit or cluster, especially main stalk bearing solitary flower or subordinate stalks

perianth the floral parts of the flower outside the sexual parts. In some flowers – but not those of proteas – these are the petals and sepals. In proteas each flower has four perianth segments in a ring around the ovary

raceme a flower cluster with the separate flowers attached by short equal flower stalks or pedicels at equal distances along a central stem

serotinous (adj.), serotiny (noun) the retention of fruits on the plant in protective cones or seed heads for long periods. In proteas this provides protection from fire and predation

scarification removal of or puncturing of a portion of the seed coat using a sharp instrument or sandpaper

scion a grafting shoot inserted into a rootstock to form new growth

stoloniferous having stolons: trailing over the soil surface and rooting at the nodes

stratification seeds stored covered in moist sand or absorbent paper at 4°C for several weeks to simulate northern hemisphere winter conditions

sub-shrub a woody perennial with a persistent underground stem producing annual shoots from ground level

unilocular having a single separate cavity/compartment in the ovary

APPENDIX 1

Summary of Germination Studies in Fynbos

Germination studies testing the effect of smoke and other factors on the seed of nearly 283 fynbos species were done at the Kirstenbosch Research Centre in Cape Town during the period 1992-2002.

Table 1 shows a summary of the proportion of species in each fynbos family that gave a germination response to smoke.

Table 1 PLANT FAMILIES TESTED FOR A GERMINATION RESPONSE TO SMOKE

✓ = Families in which at least one species studied responded to smoke
✗ = Families in which none of the species studied responded to smoke

Plant Families	Category	Total No of species tested	No of species giving significant response (%)	No of species not showing a significant response (%)
Agapanthaceae	6 ✗	1	0	1
Aizoaceae	4 ✓	21	9 (43%)	12 (57%)
Amaryllidaceae	6 ✗	1	0	1
Apiaceae	6 ✗	1	0	1
Asphodelaceae	6 ✗	2	0	2
Asteraceae	4 ✓	53	20 (38%)	33 (62%)
Boraginaceae	6 ✗	1	0	1
Brassicaceae	4 ✓	4	1 (25%)	3 (75%)
Bruniaceae	4 ✓	7	3 (43%)	4 (57%)
Campanulaceae	4 ✓	10	7 (70%)	3 (30%)
Caryophyllaceae	4 ✓	2	1	1
Crassulaceae	4 ✓	2	1	1
Cuppressaceae	6 ✗	1	0	1
Cyperaceae	6 ✗	2	0	2
Dipsacaceae	6 ✗	1	0	1
Ebenaceae	6 ✗	1	0	1
Ericaceae	4 ✓	53	33 (62%)	20 (38%)
Fabaceae	4 ✓	8	3 (38%)	5 (62%)
Gentianaceae	6 ✗	1	0	1
Geraniaceae	4 ✓	9	2 (22%)	7 (78%)
Haemodoraceae	4 ✓	4	1 (25%)	3 (75%)

Hyacinthaceae	6 ✗	2	0	2
Iridaceae	4 ✓	15	2 (13%)	13 (87%)
Juncaceae	6 ✗	1	0	1
Lanariaceae	6 ✗	1	0	1
Molluginaceae	6 ✗	1	0	1
Montiniaceae	6 ✗	1	0	1
Penaeaceae	4 ✓	4	2 (50%)	2 (50%)
Poaceae	4 ✓	3	3 (100%)	0
Polygalaceae	4 ✓	2	1	1
Proteaceae	4 ✓	44	10 (23%)	34 (77%)
Restionaceae	4 ✓	67	42(63%)	25 (37%)
Rhamnaceae	6 ✗	4	0 (0%)	4 (100%)
Rubiaceae	6 ✗	1	0	1
Rutaceae	4 ✓	5	1 (20%)	4 (80%)
Scrophulariaceae	4 ✓	13	10 (79%)	3 (21%)
Sterculiaceae	6 ✗	5	0	5 (100%)
Stilbaceae	4 ✓	1	1	0
Thymelaeaceae	4 ✓	4	3 (75%)	1 (25%)

Below: Ripe fruits of the tortoise berry, *Nylandtia spinosa* (see page 200)

APPENDIX 2

Additional List of Highly Recommended Fynbos Plants

G = recommended for garden cultivation

C = recommended for container cultivation

FYNBOS PROTEAS

Leucadendron arcuatum (G)

Leucadendron chamelaea (G)

Leucadendron comosum (G)

Leucadendron coniferum (G)

Leucadendron glaberrimum (G)

Leucadendron laxum (G)

Leucadendron modestum (G+C)

Leucadendron platyspermum (G)

Leucadendron rubrum (G)

Leucadendron spissifolium (G)

Leucadendron strobilinum (G)

Leucadendron xanthoconus (G)

Leucospermum catherinae (G)

Leucospermum conocarpodendron (G)

Leucospermum formosum (G)

Leucospermum fulgens (G)

Leucospermum lineare (G)

Leucospermum mundii (G+C)

Leucospermum patersonii (G)

Leucospermum tomentosum (G)

Protea aristata (G)

Protea aurea (G)

Protea burchellii (G)

Protea coronata (G)

Protea lacticolor (G)

Protea laurifolia (G)

Protea lepidocarpodendron (G)

Protea longifolia (G)

Protea mundii (G)

Protea nana (G+C)

Protea obtusifolia (G)

Protea punctata (G)

Serruria aitonii (G+C)

Serruria brownii (G+C)

Serruria fasciflora (G+C)

Serruria trilopha (G+C)

FYNBOS RESTIOS

Askidiosperma andreaeanum (G)

Calopsis paniculata (G)

Ceratocaryum argenteum (G)

Chondropetalum aggregatum (G)

Chondropetalum mucronatum (G)

Elegia cuspidata (G)

Elegia filacea (G+C)

Elegia racemosa (G)

Elegia stipularis (G+C)

Restio brachiatus (G)

Restio festuciformis (G)

Restio multiflorus (G)

Restio tetragonus (G)

Rhodocoma arida (G)

Rhodocoma capensis (G)

Rhodocoma fruticosa (G)

Thamnochortus acuminatus (G)

Thamnochortus cinereus (G+C)

Thamnochortus fruticosus (G)

Thamnochortus lucens (G)

Thamnochortus pellucidus (G+C)

Thamnochortus punctatus (G)

Willdenowia incurvata (G)

FYNBOS ERICAS

Erica abietina (G)

Erica blandfordia (G)

Erica caffra (G)

Erica chamissonis (G)

Erica cubica (G+C)

Erica densifolia (G)

Erica diaphana (G)

Erica formosa (G+C)

Erica glauca (G+C)

Erica gracilis (G+C)

Erica halicacaba (G+C)

Erica hirtiflora (G)

Erica holosericea (G+C)

Erica leucotrachela (G+C)

Erica longifolia (G)

Erica mammosa (G)

Erica massonii (G)

Erica mauritanica (G)

Erica nana (G+C)

Erica perspicua (G)

Erica peziza (G)

Erica plukenetii (G)

Erica porteri (G+C)

Erica propendens (G+C)

Erica sessiliflora (G)

Erica sitiens (G)

Erica taxifolia (G)

Erica thomae (G)

Erica urna-viridis (G+C)

Erica ventricosa (G+C)

Erica vestita (G)

Erica viridiflora (G)

Cannomois virgata (see page 49)

Appendix 2 contd.

FYNBOS DAISIES

Arctotis acaulis (G+C)

Arctotis aspera (G)

Chrysanthemoides incana (G)

Chrysocoma coma-aurea (G)

Cineraria saxifraga (G+C)

Dymondia margaretae (G+C)

Eriocephalus ericoides (G)

Eriocephalus racemosus (G)

Euryops speciosissimus (G)

Felicia amelloides (G+C)

Felicia filifolia (G)

Gazania rigens (G+C)

Gazania rigida (G+C)

Helichrysum petiolare (G)

Metalasia muricata (G)

Osteospermum fruticosum (G)

Phaenocoma prolifera (G)

Senecio glastifolius (G)

Stoebe plumosa (G)

Ursinia sericea (G)

FYNBOS ANNUALS

Arctotis fastuosa (G+C)

Cotula duckittiae (G+C)

Felicia dubia (G+C)

Gazania lichtensteinii (G+C)

Nemesia versicolor (G+C)

Senecio arenarius (G+C)

Steirodiscus tagetes (G+C)

Ursinia anthemoides (G+C)

Ursinia cakilefolia (G+C)

Gazania pectinata (see page 78)

FYNBOS MESEMBS

Carpobrotus acinaciformis (G)

Carpobrotus mellei (G)

Carpobrotus muirii (G)

Drosanthemum bellum (G+C)

Drosanthemum bicolor (G+C)

Drosanthemum striatum (G+C)

Lampranthus blandus (G)

Lampranthus explanatus (G+C)

Lampranthus filicaulis (G+C)

Lampranthus glaucoides (G+C)

Lampranthus multiseriatus (G+C)

Lampranthus reptans (G+C)

Lampranthus sauerae (G+C)

Lampranthus sociorum (G)

Lampranthus tegens (G)

Lampranthus variabilis (G+C)

Ruschia geminiflora (G)

Ruschia macowanii (G)

Ruschia multiflora (G)

Ruschia strubeniae (G)

FYNBOS BULBS

Agapanthus praecox (G+C)

Ammocharis longifolia (G+C)

Aristea capitata (G)

Babiana angustifolia (G+C)

Babiana disticha (G+C)

Babiana nana (G+C)

Babiana rubrocyanea (C)

Babiana stricta (G+C)

Babiana villosa (C)

Brunsvigia bosmaniae (C)

Brunsvigia orientalis (G+C)

Brunsvigia striata (C)

Bulbinella cauda-felis (G+C)

Bulbinella latifolia subsp. *doleritica* (C)

Bulbinella nutans (G+C)

Chasmanthe aethiopica (G+C)

Chasmanthe bicolor (G+C)

Cyanella orchidiformis (G+C)

Cyrtanthus angustifolius (C)

Cyrtanthus fergusoniae (C)

Cyrtanthus guthrieae (C)

Dierama pendulum (G)

Geissorhiza darlingensis (C)

Geissorhiza monanthos (C)

Geissorhiza radians (C)

Geissorhiza splendidissima (C)

Gethyllis afra (C)

Gladiolus angustus (C)

Gladiolus caeruleus (C)

Gladiolus liliaceus (C)

Gladiolus priorii (C)

Gladiolus tristis (G+C)

Gladiolus undulatus (G+C)

Haemanthus coccineus (G+C)

Haemanthus sanguineus (G+C)

Ixia curta (C)

Ixia dubia (C)

Ixia flexuosa (C)

Ixia lutea (C)

Appendix 2 contd.

Ixia maculata (G+C)

Ixia polystachya (G)

Ixia scillaris (C)

Ixia viridiflora (G+C)

Kniphofia praecox (G)

Kniphofia uvaria (G)

Lachenalia arbuthnotiae (C)

Lachenalia bulbifera (G+C)

Lachenalia fistulosa (C)

Lachenalia mathewsii (C)

Lachenalia mutabilis (C)

Lachenalia orchioides (C)

Lachenalia orthopetala (C)

Lachenalia pustulata (C)

Lachenalia rosea (C)

Lachenalia thomasiae (C)

Lachenalia viridiflora (C)

Moraea aristata (C)

Moraea comptonii (C)

Moraea gigandra (C)

Moraea insolens (C)

Moraea loubseri (C)

Moraea ochroleuca (G)

Moraea tulbaghensis (C)

Moraea villosa (C)

Nerine humilis (G+C)

Nerine sarniensis (C)

Nivenia stokoei (G)

Onixotis stricta (G+C)

Ornithogalum dubium (C)

Ornithogalum thyrsoides (G+C)

Oxalis hirta (G+C)

Oxalis obtusa (G+C)

Oxalis purpurea (G+C)

Romulea flava (G+C)

Romulea monadelpha (C)

Romulea obscura (C)

Romulea sabulosa (C)

Romulea saldanhensis (C)

Sparaxis grandiflora ssp. *acutiloba* (G+C)

Sparaxis tricolor (G+C)

Spiloxene capensis (C)

Tritonia crocata (G+C)

Tritonia deusta (G+C)

Tritonia squalida (G+C)

Veltheimia capensis (C)

Wachendorfia paniculata (G)

Wachendorfia thyrsiflora (G)

Watsonia aletroides (C)

Watsonia coccinea (G+C)

Watsonia humilis (G+C)

Watsonia marginata (G)

Watsonia meriana (G)

Watsonia spectabilis (C)

Watsonia tabularis (G)

Watsonia vanderspuyiae (G)

Zantedeschia aethiopica (G)

Opposite above left: *Ixia lutea* (see page 191); right: *Dierama pendulum* (see page 191)

Opposite below left: *Moraea ochroleuca* (see page 192); right: *Nivenia stokoei* (see page 192)

Rodney Saunders

Appendix 2 contd.

FYNBOS BUCHUS

Acmadenia obtusata (G)

Adenandra uniflora (G+C)

Agathosma apiculata (G)

Agathosma cerefolium (G)

Agathosma ciliaris (G)

Agathosma gonaquensis (G)

Agathosma imbricata (G)

Agathosma lanceolata (G+C)

Agathosma mucronulata (G)

Agathosma serpyllacea (G)

Agathosma thymifolia (G)

Coleonema album (G)

Coleonema calycinum (G)

Coleonema pulchrum (G)

FYNBOS SCROPHULARIAS

Freylinia crispa (G)

Freylinia longiflora (G)

Halleria ovata (G)

Manulea tomentosa (G+C)

Sutera caerulea (G)

Sutera cordata (G+C)

FYNBOS PELARGONIUMS

Pelargonium citronellum (G)

Pelargonium denticulatum (G)

Pelargonium echinatum (C)

Pelargonium fruticosum (G)

Pelargonium gibbosum (G+C)

Pelargonium glutinosum (G)

Pelargonium graveolens (G)

Pelargonium hispidum (G)

Pelargonium magenteum (G)

Pelargonium myrrhifolium (G+C)

Pelargonium odoratissimum (G+C)

Pelargonium panduriforme (G)

Pelargonium radens (G)

Pelargonium reniforme (G+C)

Pelargonium scabrum (G)

Pelargonium tomentosum (G)

Pelargonium tricolor (C)

Pelargonium triste (G+C)

Pelargonium zonale (G)

FYNBOS LEGUMES

Acacia karoo (G)

Erythrina caffra (G)

Lebeckia cytisoides (G)

Podalyria biflora (G)

Podalyria cuneifolia (G)

Schotia afra (G)

Sutherlandia frutescens (G)

Virgilia oroboides (G)

Opposite above: *Acmadenia obtusata* can be grown in alkaline or acid soils and is ideally suited to rock garden pockets or deep containers (see page 194)

Opposite below: *Coleonema aspalathoides* requires well-drained, acid soil (see page 128)

APPENDIX 3

List of recommended plants from some of the smaller fynbos families

ANACARDIACEAE
Rhus crenata (G)

Rhus lucida (G)

ASPHODELACEAE
Aloe arborescens (G)

Aloe ciliaris (G+C)

Aloe ferox (G)

Aloe perfoliata (G)

Aloe plicatilis (G+C)

Aloe succotrina (G)

BIGNONIACEAE
Rhigozum obovatum (G)

Tecoma capensis (G)

BORAGINACEAE
Anchusa capensis (G+C)

Lobostemon belliformis (G)

Lobostemon fruticosus (G)

Lobostemon montanus (G)

BRASSICACEAE
Brachycarpaea juncea (G)

CRASSULACEAE
Cotyledon orbiculata (green leafed variety (G+C)

Cotyledon orbiculata (grey leafed variety) (G+C)

Crassula coccinea (G+C)

Crassula ovata (G+C)

Crassula perfoliata (G+C)

Tylecodon paniculatus (G)

CUPPRESSACEAE
Widdringtonia nodiflora (G)

Widdringtonia schwarzii (G)

DIPSACACEAE
Scabiosa africana (G)

Scabiosa columbaria (G+C)

Scabiosa incisa (G+C)

Liesl van der Walt

Opposite: *Crassula ovata* is suited to rock gardens or large containers and its flowers are attractive to a wide variety of insects including butterflies (see page 196)

Left: *Scabiosa incisa* is an excellent herbaceous perennial for borders, rock gardens and containers (see page 196)

Below: *Lobostemon belliformis* forms an attractive, low, rounded shrub and can be grown in alkaline or acid soils (see page 196)

Appendix 3 contd.

GENTIANACEAE
Chironia baccifera (G+C)

Orphium frutescens (G)

LAMIACEAE
Leonotis leonurus (G)

Mentha longifolia (G+C)

Salvia africana-caerulea (G)

Salvia africana-lutea (G)

Salvia chamelaeagnea (G)

Salvia granitica (G)

Salvia muirii (G+C)

LINACEAE
Linum africanum (G)

MALVACEAE
Anisodontea scabrosa (G+C)

Hermannia pinnata (G+C)

Hermannia saccifera (G+C)

MELIACEAE
Nymania capensis (G)

Opposite: *Salvia africana-lutea* (see page 198)

Left: *Leonotis leonurus* (white form) (see page 198)

Below left: *Leonotis leonurus* (orange form) (see page 198) forms a striking backdrop to herbaceous borders or mixed fynbos beds

Below right: *Salvia muirii* is well suited to rock garden pockets, herbaceous borders and deep containers (see page 198)

Liesl van der Walt

Liesl van der Walt

Appendix 3 contd.

PLUMBAGINACEAE

Limonium perigrinum (G+C)

POLYGALACEAE

Muraltia macropetala (G)

Nylandtia scoparia (G)

Nylandtia spinosa (G)

Polygala fruticosa (G)

Polygala myrtifolia (G)

Polygala virgata (G)

PORTULACACEAE

Portulacaria afra (erect form) (G)

Portulacaria afra (sprawling form) (G+C)

Opposite above: *Limonium perigrinum* is an excellent choice for windy seaside gardens in alkaline, sandy soil (see page 200)

Opposite below: *Nylandtia spinosa* thrives in acid, sandy soil and is very waterwise (see pages 187, 200)

Left: *Muraltia macropetala* is suited to sunny rock garden pockets in gravelly or sandy soils (see page 200)

Below left: *Polygala fructicosa* likes acid, well-drained soil (see page 200)

Below right: *Polygala virgata* is an attractive, slender shrub suited to mixed fynbos beds (see page 200)

Appendix 3 contd.

RANUNCULACEAE
Knowltonia vesicatoria (G+C)

RHAMNACEAE
Phylica buxifolia (G)

Phylica ericoides (G+C)

Phylica plumosa (G)

Phylica pubescens (G)

THYMELAEACEAE
Gnidia pinifolia (G)

Gnidia squarrosa (G)

Passerina vulgaris (G)

Struthiola dodecandra (G)

Struthiola myrsinites (G)

Opposite: *Knowltonia vesicatoria* is ideal for deeply shaded conditions in humus-rich acid soils (see page 202)

Left: *Phylica pubescens* likes well-drained, acid soil in full sun (see page 202)

Below left: *Phylica plumosa* is suited to rock garden pockets in acid soil in full sun (see page 202)

Below right: *Gnidia squarrosa* is a long-lived small shrub, ideally suited to rock garden pockets or mixed fynbos beds in acid or alkaline soil in full sun (see page 202)

SOURCES OF SUPPLY OF FYNBOS SEEDS AND BULBS

By joining the Botanical Society of South Africa, South African members can take advantage of their annual catalogue of surplus seed supplied by the South African National Biodiversity Institute, (available from the seed room at Kirstenbosch) which usually has a wide selection of fynbos species on offer. The addresses of specialist fynbos seed merchants are listed in the Botanical Society's quarterly journal, *Veld & Flora*.

Membership of the Indigenous Bulb Association of South Africa (IBSA) will keep you in touch with others interested in bulbs. The Association publishes an annual bulletin and holds meetings, outings and talks.

For further information about these societies write to:

The Director
Botanical Society of South Africa
Private Bag X10
Claremont, 7735
South Africa
Tel: +27 (0) 21 797 2090/1/2/3
Fax: +27 (0) 21 797 2376
E-mail: info@botanicalsociety.org.za
Website: www.botanicalsociety.org.za

The Secretary
Indigenous Bulb Association of South Africa (IBSA)
P.O. Box 12265
N1 City, 7463
South Africa
Tel/Fax: +27 (0) 21 531 9690
E-mail: secretary@safricanbulbs.org.za
Website: www.safricanbulbs.org.za

Left: *Berzelia abrotanoides* (see page 173)

Below: *Leucadendron salignum,* male inflorescences (see page 32)

Opposite: *Ixia viridiflora* (see page 192)

Leucospermum reflexum var. *reflexum* (orange) and
Leucospermum reflexum var. *luteum* (yellow) (see page 27)

Index

Page numbers in bold denote illustrations

A

B

C

D

E

L

M

P

Q

R

S

T

U

V

W

Y

Z

Crassula coccinea (see page 196)

Phaenocoma prolifera in habitat, Elgin

Mary Duncan

Neville Brown is a specialist scientist at the Kirstenbosch Research Centre. He joined the staff of the National Botanical Gardens (later NBI, now SANBI) in 1988. His special research interest is dormancy and germination of fynbos species with particular reference to the effect of plant-derived smoke on seed germination.

He was senior lecturer in the Botany Department at the then University of Natal from 1967–1987 and assistant Dean of Science from 1984–1987. He obtained his PhD from that University in 1975 with a thesis entitled "Seed dormancy and germination in *Protea compacta* and *Leucadendron daphnoides*".

He is the author of numerous scientific papers published in local and international scientific journals. In 1998 he and Dr Jenny Jay started the Kirstenbosch Gardening Series. He is co-author of *Grow Proteas* with Deon Kotze and Philip Botha, and *Grow Restios* with Hanneke Jamieson and Philip Botha.

In 1995 he received a prize as one of the runners-up to the winner of the NBI Chairman's Award for that year.

Graham Duncan is a specialist horticulturist at Kirstenbosch Botanical Garden where he curates the collection of indigenous South African bulbs, and the display inside the Kay Bergh Bulb House of the Botanical Society Conservatory.

His numerous popular and scientific articles on bulbs have appeared in leading horticultural and botanical journals, and he is the author of several titles in the Kirstenbosch Gardening Series. In 1989 he co-authored two major publications on indigenous bulbs, *Spring and Winter Flowering Bulbs of the Cape* with Barbara Jeppe, and *Bulbous Plants of Southern Africa* with Prof. Niel du Plessis, illustrated by Elise Bodley.

His special interest in the genus *Lachenalia* resulted in the publication of a popular guide to the genus in 1988 titled *The Lachenalia Handbook*, in the *Annals of Kirstenbosch Botanic Gardens* and an MSc (*cum laude*) in Botany from the University of KwaZulu-Natal in 2005. In 2001 he was honoured with the International Bulb Society's prestigious Herbert Medal.

Above: *Mimetes cucullatus* in habitat, Elgin (see page 35)

Below: *Erica regia* in the Erica Garden at Kirstenbosch (see page 70)

Moraea aristata, a critically endangered species that thrives in cultivation (see page 192)

Botany Basics

Benefits of Foliar Feeding

Foliar Feeding or foliar fertilisation is the application of nutrients, directly onto leaves and stems where these nutrients are absorbed through microscopic structure called stomata. Foliar feeding is a supplemental means of applying macro and micronutrients, plant hormones, stimulants, amino acids and other beneficial substances. This is not a substitute for soil application but has been scientifically proven to work best in conjunction with kelp and compost that is applied as a soil application.

Plant roots systems absorb plant nutrients, but certain conditions in the soil can render these nutrients unavailable. When plants are stressed, have suffered root death or damage, are showing nutrient deficiency or being established from cuttings then foliar feeding becomes a useful method for nutrient application. Foliar feeding therefore acts as an insurance policy against plant loss

Benefits of foliar feeding with SEAGRO Organic Plant Food

- Foliar applied nutrients are more efficiently utilised by plants than soil-applied nutrients
- Foliar feeding is a triggering mechanism that stimulates the plant growth cycle
- Foliar feeding is a means to correct nutrient deficiencies in plants
- Increased yield e.g. fruit and vegetables
- Plant increased resistance to diseases and pets
- Improved drought tolerance
- Nutrients are absorbed and available to plant within 5 minutes of application where as soil applied nutrients take weeks to be available to the plant
- Foliar feeding triggers uptake of nutrients from the soil

Guidelines for effective foliar with SEAGRO Organic Plant Food

- Very dilute solutions are required. However with SEAGRO Organic Plant Food there is no chance of overdose or burning the plant
- Spray when wind is minimal especially when using a finely atomised applicator
- Nutrient uptake is increase when spray reaches the bottom surface of the leaf where the greatest concentration of stomata is found
- Spray early morning or late afternoon when stomata open. During hot days stomata closes to preserve water
- SEAGRO is compatible with most herbicides and pesticides therefore these substances can be applied together

SEA PRODUCT RANGE

SEAGRO Organic Plant Food is a nutritious emulsion containing the full range of macro and microelements, amino acids and carbohydrates in perfect balance for healthy soil and plant nutrition. When applied as a foliar feed this complex organic plant food delivers nutrients directly to the foliage for optimal plant growth and development. As a soil drench Seagro stimulates earthworm and microbial activity and replaces soil organic matter.

SEAGRO SUPERKEL is a seaweed based organic pelletised fertiliser that contains freshly harvested South African seaweed, *Ecklonia maxima* and *Laminaria pallida.* Seaweed has been used for centuries as a source of micronutrients and growth hormones, such as auxin and cytokinin. Superkel is both a fertiliser and a soil conditioner thus improving soil structure. Seaweed contains a substance called alginate that has the ability absorb large quantities of water therefore improving soil water – holding capacity.

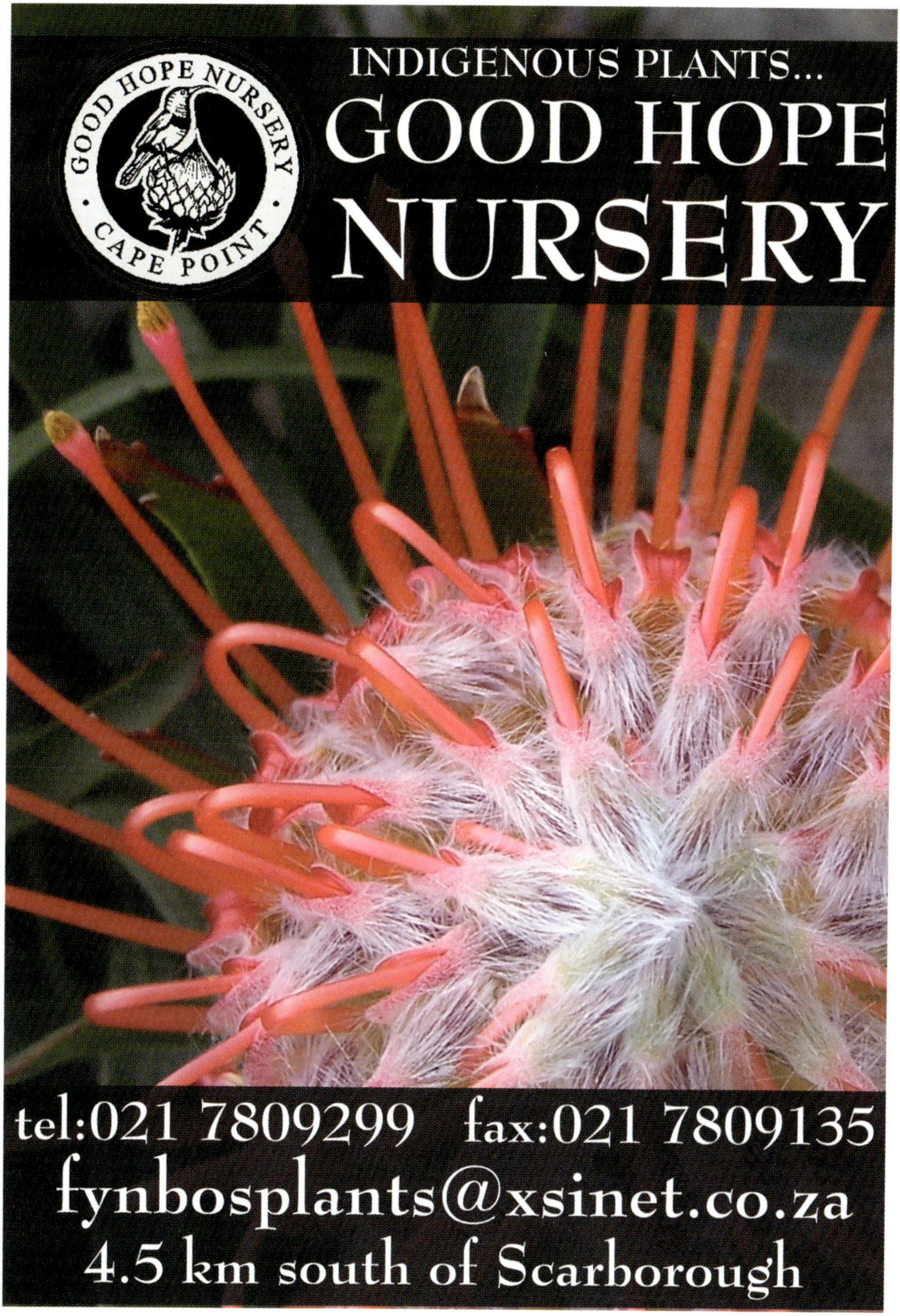
GOOD HOPE NURSERY
CAPE POINT
INDIGENOUS PLANTS...
GOOD HOPE
NURSERY
tel:021 7809299 fax:021 7809135
fynbosplants@xsinet.co.za
4.5 km south of Scarborough

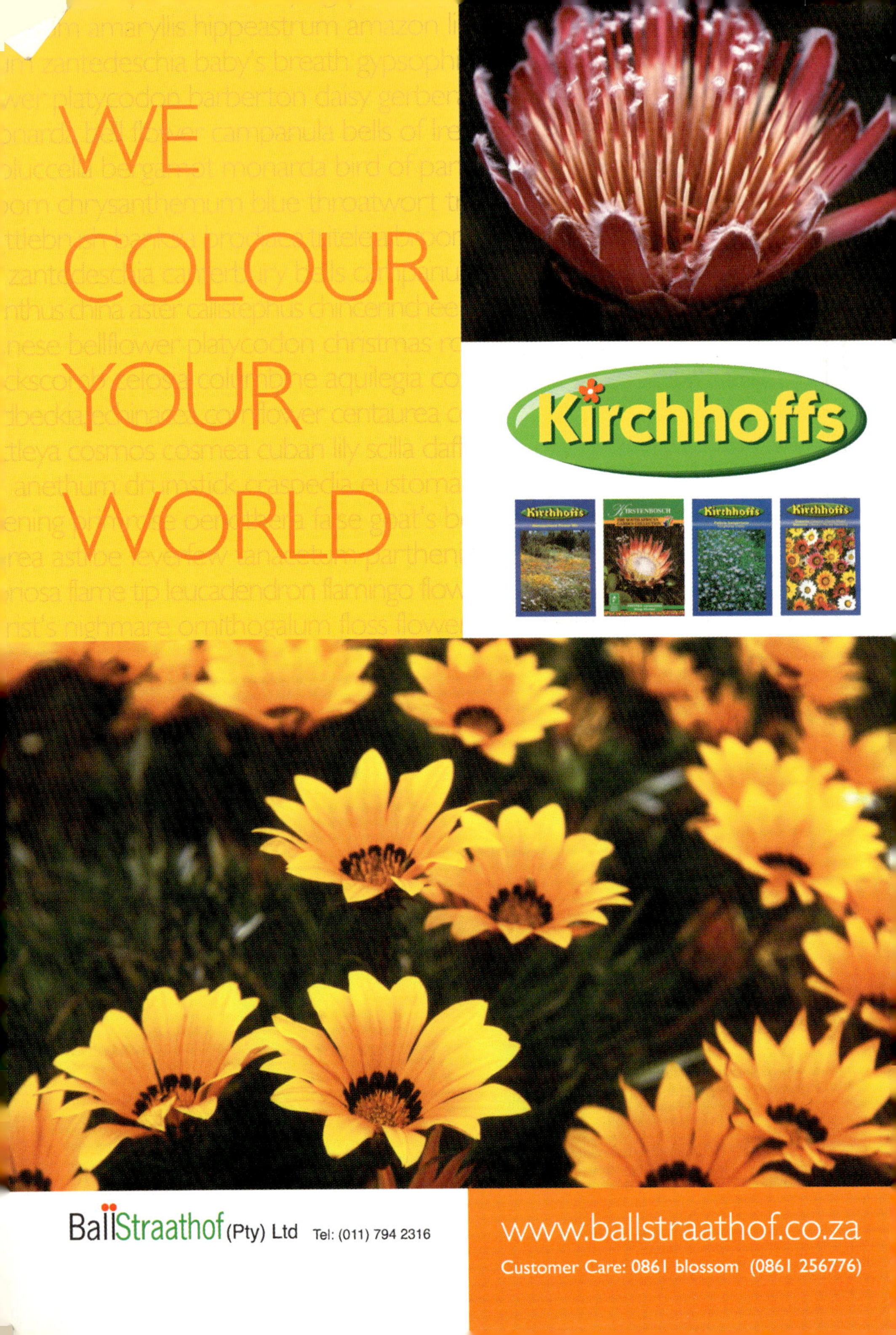
WE
COLOUR
YOUR
WORLD
Kirchhoffs
BallStraathof (Pty) Ltd Tel: (011) 794 2316
www.ballstraathof.co.za
Customer Care: 0861 blossom (0861 256776)

INDIGIGRO

Pelargonium peltatum 'Worcester' is an excellent ground cover or subject for hanging baskets (see page 161)